FENG SHUI

FOR INTELLIGENT PROCESS IMPROVEMENT

Legacy Principles of Structured Software Development

By

V. Quan Lee and J. Kay

Copyright and Publishing Notice

Editing by: Brian Loveless
Web Content Design by: Erik Victor Jones
Content Editing and Layout by: Emma Victoria Lee

ISBN: 979-8-9958725-4-2

Dedication

This restored legacy edition is dedicated to the real workers of Information Technology: the analysts, testers, developers, project managers, configuration managers, trainers, auditors, support teams, documentation keepers, and quiet process champions who carried software projects forward long before every problem had a dashboard, a collaboration tool, or a fashionable new acronym.

It is dedicated to the people who sat in conference rooms under fluorescent lights, reviewed requirements line by line, guarded baselines, chased signatures, updated process binders, corrected version numbers, and still found a way to deliver software when the process was imperfect but the responsibility was real.

Author's Note to the Restored Legacy Edition

This edition is intentionally nostalgic. It looks back at a period when structured software development was built around plans, documents, phases, gates, signoffs, audits, change boards, process libraries, and the stubborn belief that chaos could be reduced if people agreed how work should move from beginning to end.

The purpose of this restored manuscript is not to replace newer delivery approaches or to disguise legacy thinking in modern language. The purpose is to preserve the principles that made process improvement meaningful before process became a slogan, a ceremony, a maturity badge, or a consultant invoice with travel expenses attached.

At the center of the book remains the original comparison to Feng Shui: a workplace, like a living space, must be arranged so energy can flow. In software development, that energy is work, communication, accountability, knowledge, documentation, decision-making, and trust. When those forces are blocked, the workplace becomes noisy, defensive, political, and tired. When they are balanced, the workplace becomes more orderly, humane, and capable of producing quality.

How to Use This Restored Legacy Edition

This book is intentionally written as a restored legacy edition. It preserves the structured software development mindset of an earlier era: requirements, documentation, signoffs, process libraries, audits, metrics, training records, change control, and controlled implementation. It is not intended to turn the reader away from newer practices. It is intended to preserve the principles that gave software process improvement its discipline, accountability, and workplace harmony.

Readers who are new to formal software process improvement should begin with the early chapters and follow the argument in order. The book moves from the reason for process improvement, to the reality of CMMI, Agile, and SDLC, to the birth of IPI, and finally into the operating structures, forms, tools, and field guidance needed to implement it.

Readers who manage teams may use the Process Champions Committee material, the Four-Tier Process Hierarchy, and the 90-Day Legacy IPI Restoration Plan as practical implementation guidance. Readers who perform the real work may use the checklists, forms, satire, and field notes to recognize what is broken, describe it without fear, and help restore balance without becoming anti-process.

The Feng Shui comparison should be read as a practical metaphor. IPI seeks balance between process and accountability, structure and flexibility, documentation and usefulness, management oversight and worker empowerment. When the environment is arranged properly, the work can flow. When the process is arranged properly, people can breathe.

About the Authors

V. Quan Lee and J. Kay bring together two complementary views of software process improvement: the disciplined world of quality assurance, audit readiness, and formal process control, and the practical world of business analysis, project delivery, customer communication, and real-worker execution.

Their shared perspective is grounded in the belief that process improvement should not be reserved for elite methodology specialists, expensive consulting teams, or organizations large enough to finance endless training classes. Process improvement belongs to the people who must live with the process every day.

This restored edition honors that original point of view by preserving the humor, frustration, structure, and workplace realism that shaped Intelligent Process Improvement.

About the Publisher

Intelligent Systems Press publishes practical works at the intersection of software development, process improvement, technology leadership, and enterprise operating discipline.

The imprint exists to preserve and advance ideas that help organizations bring clarity to complex work. Its publishing philosophy is simple: useful books should give readers a framework, a language, a set of tools, and the confidence to improve the systems around them.

This restored legacy edition is presented as part of a broader body of work on Intelligent Process Improvement and Process Intelligence.

Other Books in the Series

Feng Shui for Intelligent Process Improvement: Legacy Principles of Structured Software Development restores the original IPI perspective and preserves the disciplined roots of software process improvement.

Process Intelligence: The Missing Link Between Agile, DevOps, and AI continues the journey by translating those enduring principles into the operating layer needed for modern software delivery organizations.

Together, the books form a series: one volume honors the old ways that taught software teams the value of structure, documentation, accountability, and balance; the next volume shows how those same principles evolve into intelligent alignment across contemporary delivery systems.

TABLE OF CONTENTS

The IPI Mission Statement

Mission Statement

The purpose of Intelligent Process Improvement is to intelligently select and implement incremental improvements that measurably enhance an organization's processes by creating harmony, clarity, accountability, and empowerment for the real workers in IT.

At the end of the day, IPI is about taking the frustration out of the software development process. This restored edition keeps that mission intact. It does not attempt to retrofit the book into the language of today's delivery movements. Instead, it returns to the older operating question: How can a software organization create enough structure, responsibility, documentation, and measurement that people can perform meaningful work without being consumed by chaos?

That question belongs to an era when the software development life cycle was not merely a chart on a wall, but the organizing backbone of the department. Requirements had a place. Design had a place. Development had a place. Testing had a place. Configuration management had a place. Metrics had a place. The purpose of structure was not to bury people under paperwork. The purpose was to create order so that work could move from concept to completion with less confusion, less blame, and less rework.

This manuscript therefore presents process improvement as a legacy discipline: practical, structured, people-aware, documentation-driven, and accountable. It recognizes that the older world of software development had limitations, but it also recognizes that the older world carried lessons that remain worth preserving: define the work, define the roles, define the artifacts, measure the result, and protect the people doing the real work.

A Note on the Legacy Spirit of This Book

This book is intentionally written as a restored legacy edition.

It does not attempt to modernize the original spirit of structured software development into the language of today's technology trends. It does not rush to replace requirements documents with user stories, phase gates with ceremonies, signoffs with informal agreement, or process libraries with whatever folder happens to be the least confusing on a shared drive.

This book honors the old ways of doing things.

Not because the old ways were perfect.

They were not.

Anyone who lived through the era of waterfall project plans, status binders, approval forms, requirements traceability matrices, change control boards, defect logs, configuration audits, and process improvement committees knows that the old ways had their own special brand of workplace comedy. A process binder could become a sacred object that everyone respected but no one opened. A status meeting could last longer than the task being reported. A document could receive ten signatures and still fail to answer the one question the developer needed answered on Tuesday afternoon.

And yet, beneath the paperwork, the meetings, the forms, the baselines, the approvals, and the occasional three-ring circus of methodology enforcement, there was an important principle worth preserving:

Software development needs order.

It needs memory.

It needs accountability.

It needs a shared understanding of how work moves from idea to requirement, from requirement to design, from design to construction, from construction to test, from test to release, and from release to maintenance.

That is the legacy spirit of this book.

The purpose of this restored edition is not to make the past look flawless. The purpose is to recover the disciplined thinking that gave structured software development its foundation. In an earlier era, process was not merely a buzzword. It was a way to create predictability in a workplace that often felt unpredictable. It was a way to give real workers a map when the project itself had become a maze.

This book keeps the nostalgic feel of that era because the nostalgia is part of the lesson.

There was a time when teams printed requirements, walked documents from cubicle to cubicle, placed signoff sheets in folders, stored templates in process libraries, argued over version numbers, reviewed defect reports in conference rooms, and waited for someone to say, "Has this been approved?" before anything important moved forward.

It was slower.

It was heavier.

It was sometimes frustrating.

But it also reminded everyone that work had consequences. Decisions had owners. Changes had impact. Documentation had value. Quality could not be assumed into existence. Someone had to define the process. Someone had to follow it. Someone had to measure it. Someone had to maintain it.

That is where Intelligent Process Improvement enters the story.

IPI is presented in this book as a practical way to bring balance and harmony into the software development workplace. It does not worship methodology for methodology's sake. It does not ask real workers to become servants of paperwork. It does not pretend that a company becomes mature simply because it has more forms than problems.

IPI asks a simpler question:

How can process improvement reduce frustration instead of creating more of it?

The Feng Shui comparison is central to that question.

Feng Shui teaches that an environment becomes healthier when energy flows properly, when objects are placed with intention, when imbalance is corrected, and when harmony is restored between people and their surroundings. This book applies that idea to the IT workplace. In software development, energy flows through requirements, communication, decisions, approvals, artifacts, metrics, defects, risks, responsibilities, and handoffs.

When that flow is blocked, frustration increases.

When requirements are unclear, confusion spreads.

When accountability is missing, finger pointing begins.

When documentation is scattered, memory disappears.

When metrics are decorative, management sees illusion instead of reality.

When processes are imposed without real-worker involvement, resistance becomes the natural response.

The purpose of IPI is to restore balance.

Balance between process and flexibility.

Balance between management oversight and worker empowerment.

Balance between documentation and actual usefulness.

Balance between measurement and common sense.

Balance between discipline and humanity.

Balance between the formal structure of software development and the everyday reality of the people who must live inside that structure.

This is why the book continues to center the "real workers" of IT. The business analysts, testers, developers, project managers, trainers, configuration managers, process librarians, auditors, support teams, and customer-facing staff are not background characters in process improvement. They are the people who know where the process breaks. They know which forms are useful and which ones exist only because someone once attended an expensive training class. They know when a meeting is solving a problem and when it is simply creating minutes for the next meeting.

A process that ignores real workers will eventually fail real work.

A process that includes real workers has a chance to become useful.

That is the practical heart of this book.

The chapters that follow preserve the structured software development mindset of a legacy process era: CMMI, SDLC, process hierarchy, governance, process libraries, audits, training, metrics, change control, lessons learned, and continuous improvement through disciplined maintenance. These concepts are presented through the lens of IPI and Feng Shui because the objective is not merely compliance. The objective is harmony.

Not soft harmony.

Disciplined harmony.

The kind of harmony where people know what is expected, where work products are controlled, where decisions are visible, where process assets can be found, where changes are evaluated before they cause damage, and where improvement is measured by whether the workplace becomes more effective — not by whether the slide deck looks more impressive.

Read this book, therefore, as both a restoration and a reminder.

It restores the legacy principles of structured software development.

It reminds us that process improvement was always supposed to improve the work, not exhaust the workers.

It reminds us that discipline and common sense should not be enemies.

It reminds us that a well-designed process can reduce chaos, protect quality, support accountability, and create a better workplace.

And, most importantly, it reminds us that every software organization — regardless of era, methodology, technology, or fashionable acronym — still needs balance.

That is the legacy spirit of this book.

How to Read This Legacy Edition

This edition should be read as a restored process-engineering manuscript, not as a modern transformation playbook. Its value is historical, practical, and cultural. It shows how structured software development tried to make order out of a chaotic technical workplace.

The language has been refined for clarity. The grammar has been polished. The structure has been strengthened. The visuals are identified for replacement with cleaner, professional, legacy-style diagrams. However, the core viewpoint remains intentionally waterfall-centric and phase-driven.

The book is nostalgic by design. It remembers a time when process engineering attempted to give software departments a common language. It remembers the importance of artifacts, process roles, committees, training, libraries, inspections, baselines, change control, and measurement. These were not bureaucratic inventions for their own sake. They were attempts to create a reliable workplace in a field that often rewarded urgency over discipline.

Readers who come from newer delivery environments may be tempted to judge every legacy practice by contemporary standards. This edition asks the reader to do something different: step into the period, understand the problem, and recognize why structure once felt like a form of relief.

Editorial Positioning for the Restored Edition

This restored edition is built around three editorial commitments.

1. Preserve the legacy concept.

The manuscript remains grounded in traditional software development, structured process improvement, SDLC alignment, CMMI-inspired discipline, role clarity, artifact control, and the language of process engineering.

2. Improve the reader experience.

The grammar, transitions, chapter openings, callouts, and visual placeholders have been refined so the book reads as a coherent professional manuscript rather than a raw process-improvement diary.

3. Retain the human-centered philosophy.

The original work was never simply about documentation. It argued that process should serve the real workers. This edition keeps that message at the center: process improvement succeeds only when people understand why the process exists and how it improves the work environment.

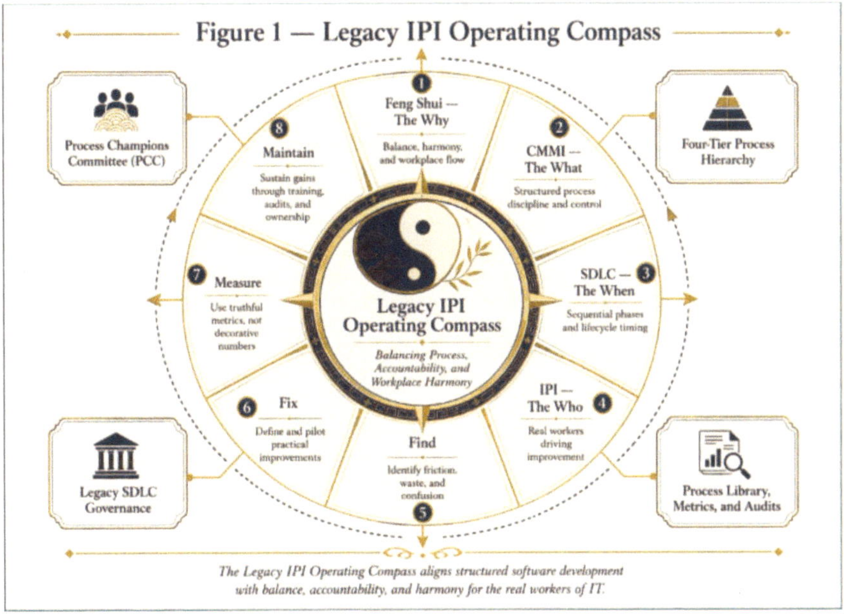

Figure 1 — Legacy IPI Operating Compass

CHAPTER 0

Feng Shui: A Reason to Process Improve

Every process improvement effort begins with a reason. Without a reason, process becomes a command. Without meaning, documentation becomes paperwork. Without visible benefit, roles become titles, reviews become interruptions, and metrics become numbers that no one trusts.

This is why the original concept placed Feng Shui before the formal discussion of software methodology. The choice was unusual, and that unusual choice is exactly what gives the book its identity. Feng Shui was not introduced as decoration. It was introduced as a metaphor for balance, placement, flow, environment, and energy inside the workplace. In a software department, the environment is not only physical. It is procedural, emotional, managerial, and cultural.

People do not experience process as theory. They experience it as meetings, approvals, templates, checklists, emails, review boards, status reports, configuration requests, audit findings, and project pressure. When these elements are poorly arranged, the workplace feels heavy. When they are arranged with purpose, the workplace begins to feel navigable. The same people who once resisted process may begin to see it as a source of stability rather than a source of aggravation.

Feng Shui, in this restored legacy context, becomes a language for explaining why order matters. A well-arranged process environment gives people room to think. It reduces ambiguity. It gives the analyst a place to capture requirements, the designer a place to shape the solution, the developer a place to build, the tester a place to verify, the configuration manager a place to control change, and the customer a place to understand progress.

In the older world of structured development, work was expected to pass through recognizable spaces. Those spaces were not merely phases on a diagram. They were work areas. Each area needed a purpose, an owner, a value, an entry condition, an exit condition, and an artifact trail. When those elements were absent, the software department became a hallway with no doors: work moved everywhere and nowhere at the same time.

The purpose of Intelligent Process Improvement is to arrange the work so that people can see where they are, what they owe, what they own, and what must happen next. This arrangement does not remove difficulty. It removes unnecessary confusion.

Legacy Principle

A process has value only when the people doing the work understand the reason for it, can perform it without unnecessary burden, and can see how it improves the quality of the workplace.

In many IT organizations, process improvement has historically been introduced as an external demand. A customer requires evidence. A contract requires compliance. An executive wants a maturity rating. A consultant arrives with binders. A methodology is announced. The workers are told to comply. The predictable result is resistance.

IPI approaches the problem from the opposite direction. It begins with the lived experience of the worker. The frustration matters. The confusion matters. The conflict between departments matters. The endless rework matters. The fatigue of unclear requirements, unstable priorities, missed deadlines, and blame-driven meetings matters. Process improvement should not ignore these realities. It should address them directly.

To restore harmony in a software environment, the organization must recognize the relationship between structure and human energy. A poorly defined process drains energy. A useful process restores energy because it gives people confidence that their work is understood and protected. The process becomes a form of environmental design.

This chapter therefore establishes the emotional and philosophical foundation for the rest of the manuscript. The later chapters will discuss CMMI, SDLC, process roles, artifacts, process hierarchy, audits, metrics, and lifecycle control. But those topics must be understood through this first lens: process exists to improve the working condition of the people who must deliver the software.

Feng Shui Is and Is Not

In this manuscript, Feng Shui is not treated as a superstition, a religion, or a decorative theme. It is used as a practical metaphor for how environments influence behavior. In IT, the environment includes the arrangement of work, the clarity of expectations, the design of communication paths, the ownership of artifacts, and the visible flow of decisions.

Feng Shui is not a substitute for software engineering. It does not replace requirements analysis, design discipline, testing, configuration management, or project control. It gives the reader a way to understand why these activities must be placed carefully inside the organization.

When the process environment is misaligned, the workers feel the consequences first. They experience late requirements, incomplete designs, unclear responsibilities, uncontrolled changes, missing artifacts, duplicate meetings, and emergency testing. When the process environment is aligned, work may still be difficult, but it is no longer needlessly disorganized.

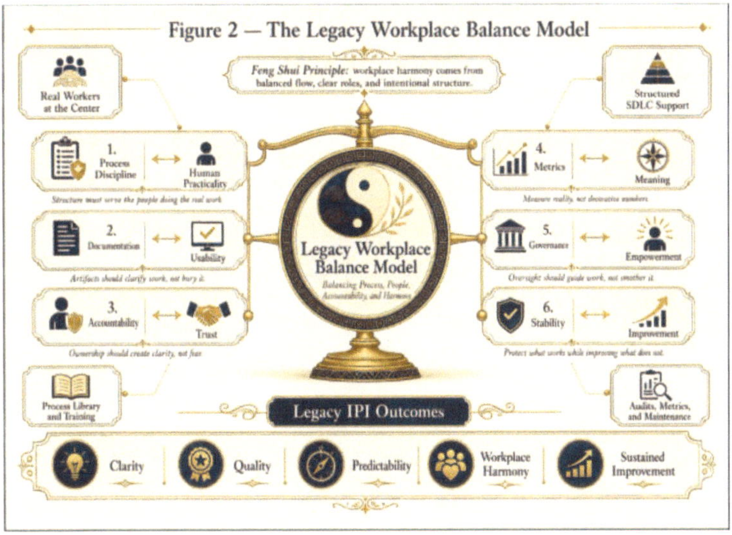

Figure 2 — The Legacy Workplace Balance Model

Why Legacy Process Improvement Still Matters

Legacy Field Observation

In many IT organizations of the legacy era, process trouble first appeared as confusion at the handoff point. People were not refusing to work; they were trying to move through a workflow that had never been properly arranged.

The practical lesson is simple: when chaotic intake combines with unrealistic promises, the organization begins paying interest on process debt. That interest appears as rework, political defense, duplicate meetings, unexplained status colors, and heroic weekends that are celebrated publicly and resented privately.

IPI treats the condition as a flow problem. The response is to locate the blockage, identify the missing process object, and restore enough structure for people to work without guessing.

Satirical Reality Check

> If the organization cannot explain invisible dependencies without opening six spreadsheets, three shared drives, two email threads, and one mysterious document named FINAL_FINAL_USE_THIS_ONE_v7, the issue is not employee motivation. The issue is process interior design. The workplace furniture may look expensive, but the workflow sofa is blocking the accountability door.

Practical Guidance

- Define the normal path for chaotic intake in one page before debating the exception path for three weeks.
- Assign a named owner for unrealistic promises, because anonymous ownership is where good processes go to become folklore.
- Create a visible artifact for invisible dependencies so teams can inspect the same evidence instead of arguing from memory.
- Measure late discovery only if the result will trigger a decision, correction, or meaningful conversation.
- Review real-worker fatigue at a fixed cadence so improvement does not depend on one exhausted champion with a color-coded binder.

Legacy Insert: What Good Looks Like

A good legacy process is visible, teachable, and useful under pressure. It helps the work move without requiring workers to translate a binder into common sense.

Control	Focus	Minimum Evidence	Review Cadence
Control Point 1	Chaotic Intake	Owner named, artifact visible, timing defined, exception path documented	Weekly during pilot; monthly after stabilization
Control Point 2	Unrealistic Promises	Owner named, artifact visible, timing defined, exception path documented	Weekly during pilot; monthly after stabilization
Control Point 3	Invisible Dependencies	Owner named, artifact visible, timing defined, exception path documented	Weekly during pilot; monthly after stabilization
Control Point 4	Late Discovery	Owner named, artifact visible, timing defined, exception path documented	Weekly during pilot; monthly after stabilization
Control Point 5	Real-Worker Fatigue	Owner named, artifact visible, timing defined, exception path documented	Weekly during pilot; monthly after stabilization

Mini-Case Scenario

A project team begins with confidence, a kickoff deck large enough to qualify as office furniture, and a schedule that assumes reality will cooperate. Two months later, the team discovers that chaotic intake was assumed, unrealistic promises was informal, and invisible dependencies was discussed in a hallway by people who are now on vacation. The project manager asks for accountability. The workers ask for documentation. The sponsor asks why the status was green last week. Everyone looks at the tester, because testers are apparently the archaeologists of bad decisions.

The IPI response is to pause the blame parade and map the actual flow. Once the blockage is visible, the team can restore ownership, artifact placement, and review discipline without turning the correction into a second project.

Chapter Checklist

#	Checklist Item	Status	Notes
1	The role accountable for the activity is named.	Yes / No / Partial	Evidence or owner
2	The required artifact is defined in plain language.	Yes / No / Partial	Evidence or owner
3	The approval point is visible before downstream work begins.	Yes / No / Partial	Evidence or owner
4	The exception path is documented and time-limited.	Yes / No / Partial	Evidence or owner
5	The measurement has an action threshold.	Yes / No / Partial	Evidence or owner
6	The process can be taught to a new employee without folklore.	Yes / No / Partial	Evidence or owner

CHAPTER 1

CMMI: The Good, the Bad, and the Reality

CMMI entered many organizations as a symbol of discipline. To some executives, it represented maturity, seriousness, and credibility. To some customers, it represented trust. To some process professionals, it represented a structured way to understand whether an organization had repeatable capabilities instead of heroic individual effort.

But to many workers, CMMI was experienced differently. It was experienced as documentation, training, audits, checklists, process area language, and a sudden increase in activities that seemed disconnected from the urgent work of building software. The gap between the promise of CMMI and the worker's lived experience is where many process improvement efforts failed.

This restored edition does not dismiss CMMI. It recognizes its value. A disciplined organization needs defined processes, managed commitments, measurement, quality assurance, configuration management, planning, tracking, and continuous improvement. These are not trivial concerns. They are the bones of a stable software organization.

The problem is not discipline. The problem is implementation without translation. A model can describe what an organization should be capable of doing, but the people still need to understand how those capabilities fit into the daily flow of SDLC work. When that translation is missing, process improvement becomes an overlay instead of an operating system.

Legacy process engineering must therefore ask a practical question: how do we make the model usable for real workers? The answer begins by connecting process expectations to the software development life cycle. Requirements activities should support requirements work.

Design controls should support design work. Configuration activities should support baselines and change control. Metrics should support management decisions and process health. Training should support role performance, not merely compliance records.

IPI Reality Check

The simplest process that works and adds value is better than an impressive process that no one can perform during real project pressure.

CMMI's strength is that it asks organizations to stop pretending that good software emerges from chaos. It acknowledges that commitments must be managed, plans must be monitored, quality must be verified, and processes must be improved. Its weakness, when badly introduced, is that it can feel abstract, generic, and difficult to map to actual work.

Workers rarely resist clarity. They resist confusion disguised as maturity. They resist process activities that appear after the work has already been done. They resist forms that no one uses. They resist metrics that do not drive decisions. They resist audits that punish rather than teach. They resist process language that makes simple work sound complicated.

An intelligent process improvement approach must respect the intent of the model while simplifying the implementation. The goal is not to create a miniature CMMI bureaucracy. The goal is to use disciplined thinking to build a practical process infrastructure that helps the organization perform.

The Good

CMMI reminds organizations that repeatability matters. A company that depends entirely on individual heroics is always at risk. When the heroic person leaves, the knowledge leaves. When a crisis occurs, the organization improvises. When a new project begins, the organization relearns what it should already know.

CMMI also reminds organizations that process must be institutionalized. A process cannot live only in the memory of the most experienced employee. It must be documented, trained, performed, measured, and improved. This is not glamorous work, but it is the work that allows organizations to grow beyond personality-driven success.

Finally, CMMI creates an important distinction between doing work and managing the capability to do work. A project may succeed once through effort. A mature organization seeks to succeed repeatedly through a defined operating structure.

The Bad

The bad emerges when organizations mistake the model for the method. A model is a reference point. It is not a day-to-day operating script. When organizations attempt to force abstract process expectations directly onto workers without tailoring, the result is frustration.

The bad also emerges when process improvement is driven by appearance. If the organization pursues process solely to satisfy an audit, win a contract, or impress a customer, the workers will sense the difference. They will know whether the process is meant to help them or merely to create evidence.

The most damaging failure occurs when process professionals become disconnected from project reality. If the process office asks for artifacts that do not align with the project lifecycle, the process becomes a separate job layered on top of the real job. IPI rejects that separation.

The Reality

The reality is that software organizations need discipline, but they need discipline that can be performed. They need documentation, but documentation must have purpose. They need audits, but audits must improve understanding. They need metrics, but metrics must guide decisions. They need configuration management, but change control must protect the product rather than merely slow the team.

The legacy lesson is not that CMMI was wrong. The lesson is that any process model must be translated into the language, rhythm, and constraints of the organization. Intelligent Process Improvement exists in that translation space.

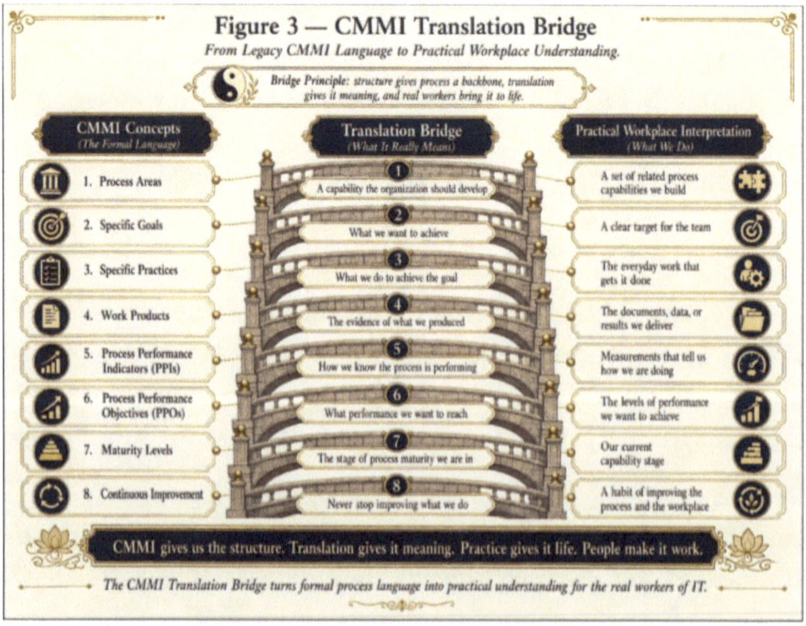

Figure 3 — CMMI Translation Bridge

CMMI Reality Notes from the Trenches

Legacy Field Observation

CMMI-related trouble often appeared when formal language reached the worker before practical translation did. The model sounded mature, but the team still needed to know what to do on Tuesday morning.

The practical lesson is simple: when audit anxiety combines with template inflation, the organization begins paying interest on process debt. That interest appears as rework, political defense, duplicate meetings, unexplained status colors, and heroic weekends that are celebrated publicly and resented privately.

IPI translates the model into practice. The response is to connect each expectation to a role, artifact, review point, and evidence trail real workers can understand.

Practical Guidance

- Define the normal path for audit anxiety in one page before debating the exception path for three weeks.
- Assign a named owner for template inflation, because anonymous ownership is where good processes go to become folklore.
- Create a visible artifact for process theater so teams can inspect the same evidence instead of arguing from memory.
- Measure rating-chasing only if the result will trigger a decision, correction, or meaningful conversation.
- Review evidence recovery at a fixed cadence so improvement does not depend on one exhausted champion with a color-coded binder.

Legacy Insert: What Good Looks Like

A good CMMI translation preserves the discipline of the model while removing the fog that makes workers feel as if maturity requires a secret language.

Control	Focus	Minimum Evidence	Review Cadence
Control Point 1	Audit Anxiety	Owner named, artifact visible, timing defined, exception path documented	Weekly during pilot; monthly after stabilization
Control Point 2	Template Inflation	Owner named, artifact visible, timing defined, exception path documented	Weekly during pilot; monthly after stabilization
Control Point 3	Process Theater	Owner named, artifact visible, timing defined, exception path documented	Weekly during pilot; monthly after stabilization
Control Point 4	Rating-Chasing	Owner named, artifact visible, timing defined, exception path documented	Weekly during pilot; monthly after stabilization
Control Point 5	Evidence Recovery	Owner named, artifact visible, timing defined, exception path documented	Weekly during pilot; monthly after stabilization

Mini-Case Scenario

A project team begins with confidence, impressive terminology, and a maturity slogan printed on the first slide. Two months later, the team discovers that audit anxiety was assumed, template inflation was informal, and process theater was discussed in a hallway by people who are now on vacation. The project manager asks for accountability. The workers ask for documentation. The sponsor asks why the status was green last week. Everyone looks at the tester, because testers are apparently the archaeologists of bad decisions.

The IPI response is to translate formal expectations into plain work instructions. The model should become usable guidance, not a vocabulary test.

Chapter Checklist

#	Checklist Item	Status	Notes
1	The role accountable for the activity is named.	Yes / No / Partial	Evidence or owner
2	The required artifact is defined in plain language.	Yes / No / Partial	Evidence or owner
3	The approval point is visible before downstream work begins.	Yes / No / Partial	Evidence or owner
4	The exception path is documented and time-limited.	Yes / No / Partial	Evidence or owner
5	The measurement has an action threshold.	Yes / No / Partial	Evidence or owner
6	The process can be taught to a new employee without folklore.	Yes / No / Partial	Evidence or owner

CHAPTER 2

Agile: Organizing Chaos Through a Legacy Lens

In this restored legacy edition, Agile is not treated as the center of the book. It is treated as a historical contrast. The original manuscript recognized Agile as a response to frustration with heavy methods, excessive documentation, and rigid controls. That recognition remains important. However, this edition intentionally preserves the waterfall-centric and process-engineering perspective.

From a legacy process viewpoint, Agile's appeal was easy to understand. People wanted relief. They wanted conversation instead of document battles. They wanted working software instead of endless planning. They wanted teams to be closer to the customer. They wanted fewer layers between decision and execution.

Yet the same organizations that needed relief also needed order. Large projects still required requirements control, architecture, integration planning, test planning, configuration management, release discipline, and accountable decision-making. The question was never whether people should communicate. The question was whether communication alone could replace a defined process infrastructure.

IPI's position in this restored edition is careful: informality can reduce unnecessary burden, but informality cannot become an excuse for unmanaged work. A software organization may desire speed, flexibility, and direct communication, but it still needs a way to know what was requested, what was approved, what was built, what was tested, what changed, and what was delivered.

The legacy value of Agile, when viewed from process engineering, is that it exposed the pain caused by processes that were too distant from real work. The legacy risk of Agile, when poorly understood, is that it may encourage organizations to dismiss structure before they have learned how to use structure intelligently.

Legacy Principle

Communication improves a process, but communication alone is not a process. A software organization still needs roles, artifacts, entry criteria, exit criteria, and change control.

The older waterfall environment often suffered because feedback arrived late. Requirements were approved, designs were completed, code was built, and testing revealed issues that should have been discovered earlier. Agile challenged that delay. It reminded software professionals that the customer's understanding evolves, that teams learn by building, and that excessive ceremony can hide poor thinking.

However, the purpose of this book is not to replace the structured life cycle with modern adaptive delivery. The purpose is to restore the legacy discipline while acknowledging why workers became frustrated with rigid implementations. The answer is not to abandon structure. The answer is to make structure intelligent.

An intelligent structured process should not require people to stop working in order to satisfy the process. It should be placed where the work naturally occurs. Requirements artifacts should help requirements conversations. Design documentation should help technical decision-making. Test plans should help quality preparation. Change requests should protect baselines. Metrics should reveal whether the process is healthy. When artifacts serve the work, workers are more likely to respect them.

The Legacy Critique of Methodology Fads

Every era has a methodology of the moment. Executives hear a term, consultants build a package, departments reorganize, workers attend training, and the organization declares a new direction. Yet the underlying problems often remain unchanged: unclear ownership, vague requirements, unstable priorities, poor handoffs, uncontrolled change, weak measurement, and insufficient accountability.

The legacy process engineer must therefore be cautious. A new vocabulary does not guarantee a new capability. If the organization cannot define work, manage commitments, control artifacts, and learn from measurement, then a new method may simply rename the old disorder.

IPI was created as a practical response to that pattern. It is not a rebellion against every method. It is a reminder that methods must serve the organization's real work. The best method is the one that creates clarity, reduces frustration, improves quality, and can be maintained by the people responsible for performing it.

Figure 4 — Informality vs. Discipline Balance

Agile Viewed Through the Legacy Lens

Legacy Field Observation

Agile-related tension appeared when necessary conversation replaced useful evidence instead of strengthening it. The team talked more, but the organization sometimes remembered less.

The practical lesson is simple: when verbal requirements combines with unrecorded decisions, the organization begins paying interest on process debt. That interest appears as rework, political defense, duplicate meetings, unexplained status colors, and heroic weekends that are celebrated publicly and resented privately.

IPI restores balance by keeping communication alive while still capturing decisions before they become folklore.

Practical Guidance

- Define the normal path for verbal requirements in one page before debating the exception path for three weeks.
- Assign a named owner for unrecorded decisions, because anonymous ownership is where good processes go to become folklore.
- Create a visible artifact for scope drift so teams can inspect the same evidence instead of arguing from memory.
- Measure hero culture only if the result will trigger a decision, correction, or meaningful conversation.
- Review calendar optimism at a fixed cadence so improvement does not depend on one exhausted champion with a color-coded binder.

Legacy Insert: What Good Looks Like

A good balance between informality and discipline allows people to speak freely while still protecting the decisions that matter.

Control	Focus	Minimum Evidence	Review Cadence
Control Point 1	Verbal Requirements	Owner named, artifact visible, timing defined, exception path documented	Weekly during pilot; monthly after stabilization
Control Point 2	Unrecorded Decisions	Owner named, artifact visible, timing defined, exception path documented	Weekly during pilot; monthly after stabilization
Control Point 3	Scope Drift	Owner named, artifact visible, timing defined, exception path documented	Weekly during pilot; monthly after stabilization
Control Point 4	Hero Culture	Owner named, artifact visible, timing defined, exception path documented	Weekly during pilot; monthly after stabilization
Control Point 5	Calendar Optimism	Owner named, artifact visible, timing defined, exception path documented	Weekly during pilot; monthly after stabilization

Mini-Case Scenario

A project team begins with confidence, enthusiastic conversation, and very few written decisions. Two months later, the team discovers that verbal requirements was assumed, unrecorded decisions was informal, and scope drift was discussed in a hallway by people who are now on vacation. The project manager asks for accountability. The workers ask for documentation. The sponsor asks why the status was green last week. Everyone looks at the tester, because testers are apparently the archaeologists of bad decisions.

The IPI response is to capture decisions at the moment of agreement. Conversation remains valuable, but memory must be protected by evidence.

Chapter Checklist

#	Checklist Item	Status	Notes
1	The role accountable for the activity is named.	Yes / No / Partial	Evidence or owner
2	The required artifact is defined in plain language.	Yes / No / Partial	Evidence or owner
3	The approval point is visible before downstream work begins.	Yes / No / Partial	Evidence or owner
4	The exception path is documented and time-limited.	Yes / No / Partial	Evidence or owner
5	The measurement has an action threshold.	Yes / No / Partial	Evidence or owner
6	The process can be taught to a new employee without folklore.	Yes / No / Partial	Evidence or owner

CHAPTER 3

Traditional SDLC: Pick and Choose

The traditional software development life cycle is one of the great organizing concepts of legacy software engineering. It gives the organization a way to understand where work begins, how it progresses, and how it reaches completion. Requirements, design, development, testing, implementation, maintenance, and retirement are more than labels. They are stations of accountability.

A common mistake is to believe that having an SDLC diagram means the organization has a defined process. An SDLC is a skeleton. It shows the broad shape of movement, but it does not provide the muscles, nerves, and connective tissue that allow the organization to perform. Process infrastructure provides that missing substance.

Each phase requires roles. Each role requires responsibilities. Each responsibility produces or reviews artifacts. Each artifact must be stored, controlled, and used. Each transition requires entry and exit expectations. Each decision point requires authority. Without these details, the SDLC remains a poster rather than a working system.

This restored edition preserves the phrase "pick and choose" because it captures an important legacy insight: not every project requires the same level of process weight. A small internal tool does not need the same artifact burden as a mission-critical enterprise system. A maintenance enhancement does not need the same governance as a new platform implementation. Intelligent process improvement means selecting the simplest set of controls that will work and add value.

The SDLC therefore becomes both structure and menu. It gives the organization a common map, while allowing project-specific tailoring. Tailoring is not the same as ignoring the process. Tailoring means consciously selecting what applies, documenting the rationale, and maintaining enough consistency that the organization can still manage quality, risk, and accountability.

Legacy Principle

The SDLC is the skeleton. Process infrastructure is the body that allows the organization to move with control, accountability, and purpose.

Core SDLC Work Areas

SDLC Work Area	Legacy Process Purpose
Initiation	Define the business need, sponsorship, scope boundaries, initial risk, and project justification.
Requirements	Document what the system must accomplish, who needs it, what constraints apply, and how requirements will be approved and traced.
Design	Translate requirements into a solution structure that can be reviewed, built, tested, and maintained.
Development	Build according to approved requirements, design expectations, coding standards, and configuration controls.
Testing	Verify that the product satisfies requirements, design intent, quality expectations, and user needs.
Implementation	Prepare the release, communicate changes, control deployment activities, and ensure readiness.
Maintenance	Operate, support, correct, enhance, and measure the system after release.
Retirement	Plan disposition, migration, archiving, decommissioning, and lessons learned.

The legacy SDLC also depends on artifacts. Artifacts are not merely documents. They are evidence of thinking. A requirements document shows that the problem has been examined. A design document shows that the solution has been considered. A test plan shows that quality has been prepared for rather than discovered by accident. A change request shows that alteration has been controlled. A lessons learned report shows that the organization is capable of memory.

When artifacts are absent, organizations rely on recollection. When recollection fails, they rely on argument. When argument fails, they rely on blame. A disciplined artifact structure reduces the need for blame because decisions, expectations, and approvals can be traced.

This is the nostalgic strength of structured development. It believed that software work should leave a trail. That trail may not have been perfect, and at times it may have become too heavy. But the underlying instinct was sound: complex work requires memory, evidence, and control.

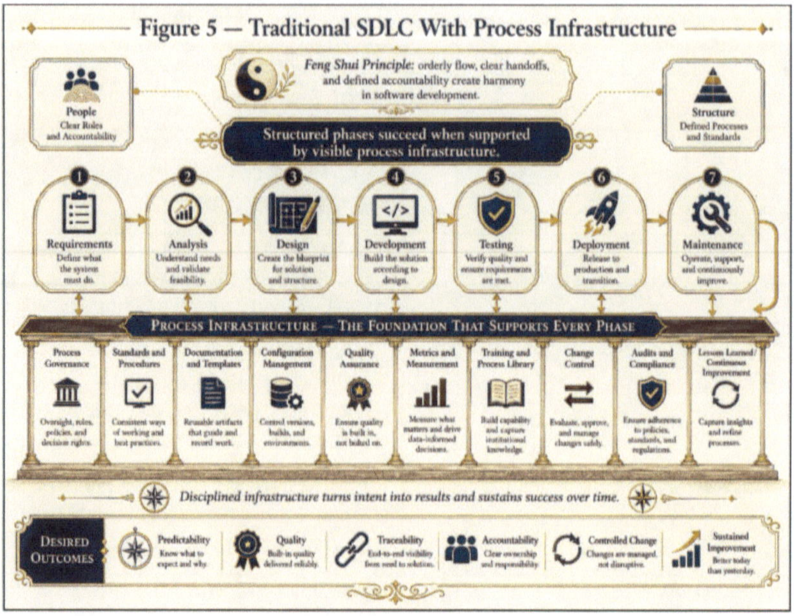

Figure 5 — Traditional SDLC With Process Infrastructure

Traditional SDLC Pick-and-Choose Guidance

Legacy Field Observation

Traditional SDLC trouble appeared when the phase diagram looked orderly but the actual handoffs were informal, late, and poorly evidenced.

The practical lesson is simple: when phase gates combines with sign-off discipline, the organization begins paying interest on process debt. That interest appears as rework, political defense, duplicate meetings, unexplained status colors, and heroic weekends that are celebrated publicly and resented privately.

IPI turns the SDLC from a wall chart into an operating structure by defining what each phase needs before it can responsibly pass work forward.

Practical Guidance

- Define the normal path for phase gates in one page before debating the exception path for three weeks.
- Assign a named owner for sign-off discipline, because anonymous ownership is where good processes go to become folklore.
- Create a visible artifact for handoff quality so teams can inspect the same evidence instead of arguing from memory.
- Measure traceability only if the result will trigger a decision, correction, or meaningful conversation.
- Review release readiness at a fixed cadence so improvement does not depend on one exhausted champion with a color-coded binder.

Legacy Insert: What Good Looks Like

A good SDLC process defines the handoff before the handoff occurs and treats evidence as protection, not punishment.

Control	Focus	Minimum Evidence	Review Cadence
Control Point 1	Phase Gates	Owner named, artifact visible, timing defined, exception path documented	Weekly during pilot; monthly after stabilization
Control Point 2	Sign-Off Discipline	Owner named, artifact visible, timing defined, exception path documented	Weekly during pilot; monthly after stabilization
Control Point 3	Handoff Quality	Owner named, artifact visible, timing defined, exception path documented	Weekly during pilot; monthly after stabilization
Control Point 4	Traceability	Owner named, artifact visible, timing defined, exception path documented	Weekly during pilot; monthly after stabilization
Control Point 5	Release Readiness	Owner named, artifact visible, timing defined, exception path documented	Weekly during pilot; monthly after stabilization

Mini-Case Scenario

A project team begins with confidence, a clean SDLC diagram, and no agreement on the actual handoff evidence. Two months later, the team discovers that phase gates was assumed, sign-off discipline was informal, and handoff quality was discussed in a hallway by people who are now on vacation. The project manager asks for accountability. The workers ask for documentation. The sponsor asks why the status was green last week. Everyone looks at the tester, because testers are apparently the archaeologists of bad decisions.

The IPI response is to define phase entry and exit conditions, then pilot them where the pain is most visible.

Chapter Checklist

#	Checklist Item	Status	Notes
1	The role accountable for the activity is named.	Yes / No / Partial	Evidence or owner
2	The required artifact is defined in plain language.	Yes / No / Partial	Evidence or owner
3	The approval point is visible before downstream work begins.	Yes / No / Partial	Evidence or owner
4	The exception path is documented and time-limited.	Yes / No / Partial	Evidence or owner
5	The measurement has an action threshold.	Yes / No / Partial	Evidence or owner
6	The process can be taught to a new employee without folklore.	Yes / No / Partial	Evidence or owner

CHAPTER 4

IPI Is Born

The Birth of IPI

Intelligent Process Improvement was born from a very old problem: organizations wanted better software delivery, but the cure often looked more painful than the disease. A company could have missed deadlines, weak requirements, informal testing, inconsistent signoffs, and defect leakage, yet the proposed remedy was often a heavy process initiative that required expensive consultants, three-ring binders, mandatory training, new acronyms, and a kickoff meeting long enough to make employees reconsider their career choices.

IPI begins with a practical observation. Most organizations do not reject process because they hate quality. They reject process because process has been presented to them as punishment. The people who do the work are told that a new methodology is arriving, that they must attend training, that their existing work is inadequate, and that a consultant who has never sat in their cubicle will now define how their work should flow. Then everyone is surprised when the real workers fold their arms, stare at the PowerPoint slides, and wait for the initiative to quietly disappear like the last four initiatives.

The restored legacy position of IPI is not anti-process. It is anti-theater. IPI believes in defined processes, documentation, audits, templates, metrics, training, signoffs, repositories, baselines, corrective actions, lessons learned, and change control. It simply refuses to pretend that these things matter if they do not help the people doing the work. A process that looks mature but makes the work environment more confused is not maturity. It is decorative bureaucracy.

IPI was created for organizations that need a quicker, more practical return from process improvement. It is for companies that cannot afford a long, disruptive implementation cycle, cannot spare months of consultant discovery, and cannot ask already overloaded workers to produce fictional as-is process documentation just so a process assessment can look complete. IPI tells the truth early: if the process is not documented, say it is not documented. Do not create a mystery novel and call it current state analysis.

This chapter restores the original idea that IPI is a workplace improvement process as much as it is a software process improvement approach. It is designed to change the atmosphere, not merely the flowchart. Feng Shui enters here not as decoration, but as the governing metaphor: a workplace out of balance produces frustration; a workplace with balanced roles, visible expectations, disciplined handoffs, and measurable accountability produces harmony.

Legacy Field Note - The Kickoff Meeting

Every legacy IT shop has seen the meeting where leadership announces a process improvement initiative with great enthusiasm while the workers silently calculate how many new templates will be added to their already impossible week. IPI begins by asking the question nobody asks in that meeting: will this make the work clearer, or will it only make the binder heavier?

The Intended Audience for IPI

IPI is intended for organizations that need process discipline but cannot survive process overkill. These organizations are often in the uncomfortable middle. They are too large to operate entirely through hallway conversations, but too practical to adopt an overbuilt model without tailoring it. They have customers, deadlines, requirements, defects, releases, support issues, and managers who need evidence. They also have workers who are tired of being told that the next methodology will finally save them.

The IPI audience includes software shops where the SDLC exists in theory but not in practice. There may be a project plan, but no one agrees on entry and exit criteria. There may be requirements, but the approval trail is buried in email archaeology. There may be testing, but test readiness is declared by optimism. There may be configuration management, but the shared drive looks like a digital attic. IPI speaks to these environments because it starts with the reality of work, not the fantasy of process maturity.

IPI is also for executives who genuinely want improvement but do not want to damage morale by imposing process as a weapon. The executive does not need to become a process engineer. The executive needs a structure that makes accountability visible, gives workers a voice, and produces evidence that the organization is improving. IPI gives leadership a way to support discipline without becoming the villain of the cubicle village.

- Companies that have outgrown informal processes but are not ready for a heavy formal model.
- Departments that need defined SDLC checkpoints, approvals, artifacts, and handoffs.
- Quality groups tired of discovering defects after everyone else has declared success.
- Business analysts who need requirements discipline without becoming paperwork clerks.
- Project managers who need honest status, realistic risk visibility, and clearer escalation.
- Configuration managers who need change control to be respected before production is on fire.
- Executives who want repeatability, accountability, and morale to improve together.

IPI Reality Check:

IPI is not for organizations seeking process decorations. It is for organizations willing to admit where the work actually breaks and then fix the break in a way people can repeat.

Not Consultants - Process Improvement Liberation Specialists

IPI does not exist to create an everlasting process priesthood. It does not want to camp inside an organization until the conference room chairs recognize the consultant by name. The IPI advisor should teach, coach, observe, help define, help pilot, help measure, and then leave the organization stronger than it was found.

Traditional consulting engagements often reward duration. The longer the discovery phase, the more impressive the invoice. The more complicated the model, the more billable the translation. The more mysterious the methodology, the more dependent the client becomes. IPI takes the opposite position. If the process cannot be explained to the people doing the work, it is not ready. If the improvement cannot be piloted by the organization, it is not owned by the organization. If the consultant must remain forever, the implementation has failed its independence test.

IPI is intended to be breezy in the old-school sense: get in, observe the environment, teach the approach, help establish the Process Champions Committee, help define the first process pilots, help measure results, and get out before becoming another permanent fixture on the budget line. The goal is not to replace the organization's judgment. The goal is to return judgment to the people who understand the work.

- Do not create client dependency where internal ownership should exist.
- Do not confuse a long engagement with a meaningful engagement.

- Do not produce deliverables no one can maintain after the consultant leaves.
- Do not use methodology language as camouflage for lack of practical progress.
- Do not force the company into a one-size-fits-all implementation simply because the template is already built.

Legacy Field Note - The Eternal Engagement

Some consulting efforts look less like process improvement and more like a timeshare presentation with a project code. Every month has a new assessment, a new slide deck, and a new reason the organization is not quite ready to proceed. IPI rejects that treadmill. Process improvement should not require a subscription to confusion.

The IPI Holistic View

IPI takes a wholistic view of process improvement: Intelligence, Surroundings, and Values. Intelligence means the organization should understand its own work before changing it. Surroundings reflect the Yin Yang balance between process and accountability, structure and flexibility, management needs and worker reality. Values reflect the Feng Shui principles that turn process into workplace harmony: unity, balance, prosperity, reputation, partnership, creativity, progress, promotion, longevity, health, and wholistic learning.

Traditional process improvement often names people, process, and technology. IPI keeps those but asks a more direct workplace question: are the people who define the process, the people who execute the process, and the people who benefit from the process in alignment? When they are not aligned, negative Qi appears in ordinary IT forms: vague requirements, defensive meetings, surprise defects, production panic, unclear ownership, passive resistance, and the sacred phrase heard in many cubicles: "I thought someone else was handling that."

The purpose of IPI is not to make the organization mystical. The purpose is to make imbalance visible. Feng Shui gives IPI its language of harmony, but IPI turns that language into implementation discipline. Harmony is not merely a feeling. In IPI, harmony must be designed, trained, documented, measured, and maintained.

- Intelligence asks: do we understand the problem before selecting the process remedy?
- Surroundings ask: does the physical, organizational, and technical environment support the process?
- Values ask: does the process promote unity, balance, partnership, creativity, progress, and reduced frustration?
- Accountability asks: can the organization prove that the process is followed and effective?
- Harmony asks: are real workers more capable and less frustrated because of the process?

The Four-Part IPI Solution Set

The IPI Solution Set restores order through four practical components. These components are intentionally simple because the real work is already complicated. A process improvement model does not have to become another software system the organization must debug. It should create a stable structure that real workers can learn, use, and improve.

- Process Champions Committee (PCC): the governing and representative body that defines, pilots, communicates, and sustains process improvement.
- Bottom-Up Four-Tier Process Hierarchy: the structure that moves from real worker experience to department practice, organizational policy, and executive visibility.
- Find, Fix, Measure, Maintain (FFMM): the lifecycle for identifying process problems, creating practical corrections, measuring results, and preventing backslide.
- Legacy IPI SDLC: the disciplined software development flow that uses phases, artifacts, reviews, signoffs, and controlled transitions.

In the restored legacy concept, these components are not modernized into automation pipelines or abstract operating systems. They remain deliberately old-school. They are about committees, charters, process libraries, artifact registers, approval trails, meeting minutes, review checklists, training logs, audit evidence, metrics baselines, and lessons learned. This is the world where process discipline lived before every tool promised to solve what people had not agreed to define.

IPI Reality Check:

A tool can store a process. It cannot create process discipline. A dashboard can show a number. It cannot decide whether the number means progress or theater. IPI begins with human accountability, then uses tools as support.

The IPI Value Matrix

Each IPI component must be examined through the values of intelligent selection, incremental change, measurable results, harmony, empowerment, real-worker involvement, and frustration reduction.

- Intelligently: choose the process improvement that solves the actual problem, not the one that looks fashionable in a management seminar.
- Incremental: implement improvements in practical steps so the organization can absorb change without creating process indigestion.
- Measurably: define evidence so improvement can be observed rather than merely announced.
- Harmony: ensure the process reduces confusion, clarifies expectations, and supports cooperation.
- Empowerment: give real workers authority to define what they must execute.
- Real Workers: keep the people who live with the process at the center of process design.

- Frustration: identify, reduce, and monitor the workplace friction that process improvement is supposed to relieve.

This matrix prevents IPI from becoming another methodology label. It forces every process decision to answer a practical question: will this help the work become clearer, calmer, more accountable, and more repeatable? If not, the process may be technically correct but operationally useless.

Intelligently

Intelligent process improvement begins with selection. The organization must resist the urge to improve everything at once. In legacy IT, this temptation often appears after an executive meeting where someone says the department needs maturity, and suddenly every form, checklist, signoff, procedure, and template becomes urgent. IPI says no. Start by selecting the improvement that matters most to the people and projects currently suffering.

Intelligence also means identifying root causes instead of chasing visible symptoms. A missed deadline may look like a project management problem, but the root may be late requirements approval, weak change control, unclear test entry criteria, or unrealistic executive commitments made before the work was estimated. A defect spike may look like a testing problem, but the root may be incomplete design review or uncontrolled requirement changes. IPI restores the old-school discipline of asking what happened, why it happened, where it happened in the SDLC, and who was affected before prescribing a process cure.

An intelligent IPI improvement is small enough to pilot, clear enough to explain, and important enough to matter. It should have a visible owner, a defined artifact, an expected behavior, an approval point, and a way to measure whether it helped. If any of those pieces are missing, the improvement is not yet intelligent; it is only an idea with a heading.

Intelligently - Legacy Checklist

- What specific problem is being solved?
- Where in the SDLC does the problem appear?

- Which real workers experience the pain?
- What artifact or decision point would reduce ambiguity?
- What evidence will show that the improvement worked?

IPI Reality Check:

Intelligently must be visible in the process artifact, not merely claimed in the process philosophy.

Incremental

IPI is incremental because organizations do not become disciplined by memo. They become disciplined through repeated, visible practice. A company that has lived for years with informal handoffs cannot become fully process mature because a new binder appears on Monday. The binder may be impressive. The employees may even nod during training. But by Wednesday afternoon, the project is late, the customer is calling, and the old ways quietly return through the side door.

Incremental implementation protects the organization from process shock. It allows one workgroup to pilot a requirements checklist before every department is required to use it. It allows a test readiness review to be tried on one project before it becomes policy. It allows a change request form to be adjusted before people create unofficial shortcuts. The old ways of doing things were not always wrong; many of them emerged because people were trying to survive. IPI respects that history while replacing survival habits with repeatable discipline.

The incremental approach also aligns with Feng Shui balance. Too little change leaves the workplace in chaos. Too much change creates resistance. Balanced change moves the organization forward without overwhelming the people who must carry the improvement.

Incremental - Legacy Checklist

- Pilot before mandating.
- Improve one high-friction area at a time.

- Collect lessons learned before .
- Let the process prove itself in real project conditions.
- Do not punish people for struggling with a new process that has not yet been stabilized.

IPI Reality Check:

Incremental must be visible in the process artifact, not merely claimed in the process philosophy.

Measurably

Measurement is the moment process improvement must confront reality. IPI does not measure only the usual project quantities of time, money, defects, and productivity. Those matter, but they do not tell the whole workplace story. IPI also measures whether the process is reducing frustration, increasing clarity, improving handoffs, strengthening accountability, and supporting harmony.

In the old ways of doing things, metrics were often treated like ceremonial objects. They were placed on status reports, shown on slides, and discussed with serious faces. Then everyone returned to work without changing behavior. IPI rejects decorative numbers. A metric must either help the organization decide, learn, correct, or predict. If it does none of those things, it is not a metric; it is office wallpaper.

Measuring harmony may sound unusual until the organization asks what disharmony costs. How many defects were caused by unclear ownership? How many rework cycles came from misunderstood requirements? How many escalations came from poor communication? How many sick days, resignations, or silent disengagements came from process frustration? IPI insists that if frustration affects delivery, frustration deserves measurement.

Measurably - Legacy Checklist

- Baseline before the process change.

- Measure the same behavior after the process change.
- Separate improvement evidence from presentation decoration.
- Track whether positive change is sustained.
- Use surveys, interviews, observations, audits, defect trends, rework counts, and lessons learned together.

IPI Reality Check:

Measurably must be visible in the process artifact, not merely claimed in the process philosophy.

Harmony

Harmony is the heart of the Feng Shui comparison. In IPI, harmony does not mean everyone is happy every minute. Software development is not a day spa with requirements. Harmony means roles are understood, handoffs are visible, accountability is balanced, and the work environment supports purposeful effort rather than constant aggravation.

A harmonious IT workplace still has deadlines, defects, disagreements, and pressure. The difference is that the organization has a way to process those tensions without turning them into personal conflict. A requirement change becomes a controlled change request, not a hallway surprise. A defect becomes evidence for improvement, not a weapon for blame. A missed milestone becomes a risk analysis, not a guessing contest. Harmony is created when the process gives people a constructive path through conflict.

Feng Shui teaches that environment affects human life. IPI applies that principle to the software development environment. A cluttered repository, unclear SDLC, invisible approvals, and chaotic meetings create negative energy because they make the work harder than it has to be. Process discipline arranges the workplace so effort can flow.

Harmony - Legacy Checklist

- Are process roles clearly defined?
- Are handoffs visible and documented?
- Are decisions captured where people can find them?
- Are meetings producing outcomes or only calendar fatigue?
- Are real workers less confused after the process is implemented?

IPI Reality Check:

Harmony must be visible in the process artifact, not merely claimed in the process philosophy.

Empowerment

Empowerment in IPI is not a motivational poster. It is authority tied to responsibility. The people who do the real work must have a voice in defining the process they are expected to follow. Otherwise the process becomes something done to them instead of something built with them.

Old-school organizations often confused empowerment with permission to work harder. A worker could be empowered to meet impossible deadlines, empowered to fix defects created upstream, empowered to attend extra meetings, and empowered to take ownership of problems they did not create. IPI restores the word to its useful meaning: workers should be empowered to identify gaps, propose fixes, pilot improvements, challenge unusable procedures, and measure whether the new approach improves reality.

Empowerment must also be balanced with accountability. A workgroup cannot demand authority and then reject evidence. A Process Champions Committee cannot define a process and then ignore noncompliance. Executive leadership cannot ask for ownership and then override the process whenever urgency appears. IPI empowerment is reciprocal: voice, responsibility, evidence, and correction must travel together.

Empowerment - Legacy Checklist

- Give workgroups representation through the PCC.
- Let process users validate process usability.
- Require owners for every process artifact.
- Create escalation paths that do not punish honesty.
- Recognize compliance and practical improvement as professional contributions.

IPI Reality Check:

Empowerment must be visible in the process artifact, not merely claimed in the process philosophy.

Real Workers

The phrase real workers remains essential because IPI was never designed for methodology spectators. Real workers are the analysts, testers, developers, project managers, configuration managers, trainers, auditors, support staff, and subject matter experts who live with the consequences of process decisions. They know where the work breaks because they are usually standing there when it breaks.

Real workers are often asked to provide evidence after the fact. They are asked to reconstruct decisions, locate approvals, explain scope changes, justify defects, and defend schedules that were never realistic. IPI changes their role from evidence scavenger to process participant. They help define what must be captured, when it must be captured, where it must be stored, and how it must be reviewed.

This does not mean every worker gets a veto over every process. It means the process cannot be considered complete until the people expected to follow it have tested it against actual work. A process designed without real-worker input often looks clean because it has never been touched by reality.

Real Workers - Legacy Checklist

- Ask real workers where the process breaks.
- Document the informal workaround before replacing it.
- Pilot the process with people who will actually use it.
- Treat resistance as information before treating it as defiance.
- Use lessons learned to refine the process rather than blame the workers.

IPI Reality Check:

> Real Workers must be visible in the process artifact, not merely claimed in the process philosophy.

Frustration

Frustration is not a side issue in IPI. It is one of the main signals that the work environment is out of balance. IT frustration appears when deadlines change without process control, when requirements are vague, when signoffs are missing, when tools do not match procedures, when meetings do not produce decisions, when people are blamed for gaps they warned about weeks earlier, and when management asks for maturity without giving people time to mature the process.

IPI does not promise to remove all frustration. That would be dishonest, and dishonesty is already overrepresented in many project status reports. IPI promises to identify preventable frustration and reduce it through structured process principles. Some stress belongs to the nature of software development. But chaos caused by unclear ownership, missing artifacts, uncontrolled changes, and process inconsistency is not destiny. It is a fixable condition.

The Feng Shui comparison matters here. Negative energy in the workplace does not always announce itself with dramatic conflict. Sometimes it looks like silence in meetings, late responses, passive compliance, sick-day patterns, hallway sarcasm, or employees who stop offering ideas because they know the process will ignore them. IPI treats those signals as workplace data.

Frustration - Legacy Checklist

- Ask what frustration looks like in the workplace.
- Identify which process gap creates the frustration.
- Measure whether the corrective process reduces the frustration.
- Make frustration reduction part of the process objective.
- Do not confuse quiet compliance with harmony.

IPI Reality Check:

Frustration must be visible in the process artifact, not merely claimed in the process philosophy.

Problem Solving with IPI

In legacy IT, the list is familiar. Deadlines are dictated before estimates are understood. Software quality is compromised because testing becomes the schedule reserve. Management wants success at all costs, especially if the cost is employee morale. Requirements are complex, customers are uninvolved, resources are distributed, programmers are rushed, and the project plan becomes less a plan than a public expression of hope.

IPI does not begin by pretending these problems are unusual. It begins by admitting that many organizations experience all of them at once. The issue is not that workers cannot name the problems. The issue is that they do not know where to begin without becoming overwhelmed. IPI says start somewhere real. Find a specific process problem. Fix it in a practical way. Measure whether it helped. Maintain the improvement long enough for it to become normal.

Do not panic, do not overanalyze, and do not wait for the perfect enterprise process architecture before improving the handoff that is hurting the current project. If test entry criteria are unclear, fix test entry criteria. If requirements approval is vague, fix requirements approval. If change control exists only when someone remembers it, fix change control. The point is not to create a perfect process universe by

Friday. The point is to prove that disciplined improvement can restore order one process area at a time.

- Market pressure and imminent deadline pressure.
- Poor software quality and late defect discovery.
- Complex requirements and uncontrolled changes.
- Uninvolved or unavailable customers.
- Distributed teams and fragmented communication.
- Sloppy programming practices and uneven skill sets.
- Big design documents that are approved but not understood.
- Project managers afraid to tell leadership bad news.
- Subjective project status and decorative green indicators.
- Rework caused by vague decisions and missing evidence.
- Methodologies practiced with no visible return on investment.
- Expensive technology explored before basic process discipline exists.

Processes Do Not Fix Problems - People Do

One of the most important restored principles is that processes do not fix problems by themselves. People define processes, follow processes, audit processes, adjust processes, and decide whether processes remain alive. A documented procedure cannot walk into a meeting and demand clarity. A checklist cannot insist on honest status. A template cannot stop an executive from promising a date before the team estimates the work. Process gives people a structure, but people must bring courage and accountability to the structure.

This is why IPI insists on the Process Champions Committee. Without a governing body, process becomes scattered. Without representation, process becomes resented. Without metrics, process becomes opinion. Without training, process becomes rumor. Without audits, process becomes optional. Without lessons learned, process becomes forgetful. The PCC is the old-school sheriff in the process town, not because it bullies workers, but because it gives the organization a place where process ownership is visible.

- Processes cannot stop change, but they can control its impact.
- Processes cannot create funding, but they can show the cost of underfunding.
- Processes cannot make every employee equally skilled, but they can create consistent expectations.
- Processes cannot make everyone comply, but they can identify patterns of noncompliance.
- Processes cannot remove all conflict, but they can give conflict a constructive path.
- Processes cannot create harmony unless they are built with harmony as a measurable objective.

IPI Reality Check:

A process is a map. It cannot move the traveler. But without the map, every project tells itself it is exploring new territory when it is really lost in the same woods as last time.

The Legacy Process Language

Process improvement has its own language, and the restored Chapter 4 keeps that language proudly visible. This is the vocabulary of the old ways: plans, audits, records, controls, checklists, guidelines, templates, guidance, monitoring, artifacts, proof, reviews, entrance criteria, exit criteria, traceability, quantifying, qualifying, baselines, waivers, approvals, repositories, corrective actions, and lessons learned.

The point is not to bury workers in vocabulary. The point is to give the organization a shared language for disciplined work. When one person says signoff and another person hears casual agreement, the process is not aligned. When one department says complete and another department asks where the evidence is, the process is not aligned.

When leadership asks for status and receives adjectives instead of artifacts, the process is not aligned. IPI restores shared meaning.

- Plan means the work has an intended sequence, owner, and expected evidence.
- Audit means the organization checks whether the process is actually followed.
- Record means a decision or result is captured for future reference.
- Control means change is reviewed before it damages downstream work.
- Checklist means recurring expectations are made visible.
- Template means workers do not recreate structure from memory every time.
- Traceability means the organization can connect requirements, design, development, testing, and release evidence.
- Lessons learned means experience is converted into organizational memory.

Legacy Field Note - The New Sheriff

Once IPI is implemented, the organization can finally say there is a new sheriff in town. Its name is PCC. It carries a charter, not a six-shooter. It enforces process with evidence, not swagger. And unlike the old sheriff, it files the meeting minutes where people can find them.

IPI Is Born as a Worker-Centered Alternative

Legacy Field Observation

IPI trouble appeared when everyone agreed improvement was needed, but nobody wanted another initiative that confused participation with paperwork.

The practical lesson is simple: when bottom-up insight combines with pilot first, the organization begins paying interest on process debt. That interest appears as rework, political defense, duplicate meetings, unexplained status colors, and heroic weekends that are celebrated publicly and resented privately.

IPI begins with usefulness. If the improvement does not reduce frustration, clarify work, or protect quality, it deserves redesign before rollout.

Practical Guidance

- Define the normal path for bottom-up insight in one page before debating the exception path for three weeks.
- Assign a named owner for pilot first, because anonymous ownership is where good processes go to become folklore.
- Create a visible artifact for plain language so teams can inspect the same evidence instead of arguing from memory.
- Measure buy-in only if the result will trigger a decision, correction, or meaningful conversation.
- Review practical adoption at a fixed cadence so improvement does not depend on one exhausted champion with a color-coded binder.

Legacy Insert: What Good Looks Like

A good IPI implementation begins with a problem real workers recognize and ends with a process they can actually perform.

Control	Focus	Minimum Evidence	Review Cadence
Control Point 1	Bottom-Up Insight	Owner named, artifact visible, timing defined, exception path documented	Weekly during pilot; monthly after stabilization
Control Point 2	Pilot First	Owner named, artifact visible, timing defined, exception path documented	Weekly during pilot; monthly after stabilization
Control Point 3	Plain Language	Owner named, artifact visible, timing defined, exception path documented	Weekly during pilot; monthly after stabilization
Control Point 4	Buy-In	Owner named, artifact visible, timing defined, exception path documented	Weekly during pilot; monthly after stabilization
Control Point 5	Practical Adoption	Owner named, artifact visible, timing defined, exception path documented	Weekly during pilot; monthly after stabilization

Mini-Case Scenario

A project team begins with confidence and a promise that this process initiative will be different from the last one. Two months later, the team discovers that bottom-up insight was assumed, pilot first was informal, and plain language was discussed in a hallway by people who are now on vacation. The project manager asks for accountability. The workers ask for documentation. The sponsor asks why the status was green last week. Everyone looks at the tester, because testers are apparently the archaeologists of bad decisions.

The IPI response is to start with one improvement that workers believe will help, then prove it through use rather than speeches.

Chapter Checklist

#	Checklist Item	Status	Notes
1	The role accountable for the activity is named.	Yes / No / Partial	Evidence or owner
2	The required artifact is defined in plain language.	Yes / No / Partial	Evidence or owner
3	The approval point is visible before downstream work begins.	Yes / No / Partial	Evidence or owner
4	The exception path is documented and time-limited.	Yes / No / Partial	Evidence or owner
5	The measurement has an action threshold.	Yes / No / Partial	Evidence or owner
6	The process can be taught to a new employee without folklore.	Yes / No / Partial	Evidence or owner

CHAPTER 5

The IPI Process Champions Committee

The Process Champions Committee is the governing body of Intelligent Process Improvement. In a legacy software department, nothing meaningful survives on enthusiasm alone. Enthusiasm announces the initiative. Structure sustains it. The PCC exists because someone must protect the process after the kickoff meeting ends, after the pastries disappear, after the CIO moves on to the next emergency, and after the consultants have submitted an invoice that requires its own approval workflow.

The central idea is simple: process improvement must be owned by the people who understand the work, supported by management, and disciplined enough to survive normal project pressure. A process that only exists while an executive is watching is not a process. It is a performance. A process that only exists inside a consultant binder is not a process either. It is office furniture with a table of contents.

The PCC gives the organization a practical forum for defining, reviewing, piloting, approving, training, auditing, and improving the processes that govern software work. Its purpose is not to create a bureaucracy above the workers. Its purpose is to keep process ownership close enough to the real work that the process remains useful.

IPI Reality Check

A process initiative without a PCC is like a software release without configuration control: everyone feels busy, everyone feels important, and no one can explain which version went to production.

Why the PCC Exists

Every organization eventually learns that undocumented good intentions do not scale. One team may have a brilliant way of handling requirements. Another team may have a disciplined test signoff checklist. Another team may know how to keep configuration chaos from becoming a full-contact sport. Yet without a mechanism for capturing and institutionalizing these practices, the organization relearns the same lessons project after project.

The PCC exists to prevent the organization from becoming dependent on memory, heroics, hallway agreements, and the one person who knows where the real template is stored. In a legacy process environment, knowledge must be converted into usable artifacts. Practices must become repeatable. Roles must be defined. Training must be delivered. Compliance must be inspected. Metrics must be reviewed. Management must be informed.

This does not mean every process needs to become a 40-page policy with its own glossary. It means that every critical activity needs enough definition that the work can be performed consistently by competent people without requiring psychic powers, emergency meetings, or a ceremonial hunt through old email attachments.

The PCC as a Balanced Process Body

The PCC should not be a dictatorship, a suggestion box, or a weekly meeting created because someone needed to add governance to a slide. It should be a balanced decision-making body with representation from the functions that carry the SDLC forward. The committee must include people who understand the reality of the work and people who can authorize the changes required to improve it.

Standing Members and Rotating Representatives

The PCC should include standing members who maintain continuity and rotating representatives who keep the committee connected to the real work. Standing members provide stability. Rotating members provide reality. One without the other creates imbalance.

PCC Function	Legacy Purpose	Feng Shui Value
Executive Management	Sponsor the initiative, remove obstacles, and confirm that process improvement has organizational priority.	Unity
Customer or Business Representative	Provide visibility into expectations, acceptance, and business constraints.	Balance
Rotating Workgroup Representative	Bring the voice of real workers into process definition and pilot feedback.	Prosperity
Process Trainer	Teach new and revised procedures so adoption is not based on rumor.	Reputation
Process Librarian	Control process assets, templates, standards, and reference materials.	Partnership
Process Project Manager	Plan and coordinate the process improvement initiative as real work.	Creativity
Process Business Analyst	Map current and future process states and document the gaps.	Progress
Process Auditor	Inspect whether approved processes are being followed and whether they make sense.	Promotion
Configuration Manager	Protect baselines, versions, changes, and release discipline.	Longevity
Metrics Coordinator	Collect and interpret measures that show whether improvement is real.	Health
IPI Advisor	Coach, mentor, and keep the initiative aligned with the people-centered mission.	Wholistic

If the committee is made only of permanent process specialists, it risks drifting away from the workplace and becoming a small kingdom of templates. If it is made only of temporary volunteers, it risks losing continuity and repeating old debates every quarter. The IPI answer is to combine both: a stable process infrastructure with rotating workgroup participation.

Rotating representation also reduces the dangerous belief that process belongs to someone else. When every function periodically serves on the PCC, the organization learns that process improvement is not a mysterious discipline practiced by people with laminated badges. It is civic responsibility inside the software department.

IPI Reality Check

Serving on the PCC should feel like jury duty for process improvement: not always convenient, occasionally uncomfortable, but necessary if the department wants justice for requirements, testing, documentation, and the poor release manager who keeps inheriting everyone else's emergency.

The PCC Operating Rhythm

A PCC should operate on a defined rhythm. Legacy organizations often failed because process improvement was either too casual or too ceremonial. The casual version sounded like, "Let us all try to do better." The ceremonial version involved a kickoff, seven binders, a catered lunch, and no measurable change. The PCC needs a rhythm that turns intent into controlled action.

Activity	Recommended Cadence	Output
Process issue intake	Weekly or biweekly	Log of problems, proposals, and owner assignments
Current-state review	As needed by process area	Documented process map or current-state statement
Gap analysis	Per selected improvement	Gap list, severity, root cause, and affected roles
Process definition workshop	Per improvement cycle	Draft procedure, templates, artifact list, and role responsibilities
Pilot review	At pilot midpoint and close	Pilot findings, adoption issues, metrics, and worker feedback
Approval meeting	Monthly or per release window	Approved process, deferred process, or rejected process decision
Training rollout	Before enforcement	Training materials, attendance record, and knowledge check
Audit or assessment	After adoption period	Compliance results, improvement recommendations, and corrective actions

PCC Charter

The PCC should begin with a charter that defines authority, scope, membership, responsibilities, decision rights, and escalation paths. Without a charter, the committee may become a discussion club. Discussion clubs are pleasant, especially if coffee is provided, but they rarely fix software delivery.

The charter should state that the PCC has authority to recommend, pilot, approve, revise, retire, and communicate process assets within its assigned scope. It should also state that the committee serves the organization, not itself. The PCC does not exist to create process theater. It exists to reduce confusion, standardize important work, and improve the conditions under which software is delivered.

The charter should also protect the committee from becoming the dumping ground for every organizational complaint. The PCC is not human resources, not facilities, not therapy, not procurement, and not the place to relitigate why the conference room projector has required "just one more adapter" since 2004. Its scope is process improvement.

PCC Decision Rules

A process decision should be based on four questions: Does the proposed process solve a real problem? Can the real workers perform it without unreasonable burden? Does it create a useful artifact or control point? Can its effectiveness be measured? If the answer to these questions is unclear, the process should not be institutionalized yet.

Decision Question	What It Protects Against
What problem does this process solve?	Creating procedures because a methodology book said so.
Who performs the work?	Assigning responsibility to imaginary resources or overloaded teams.
What artifact proves the work was done?	Processes that vanish when audited.
What is the entry and exit condition?	Work that begins too early, ends too late, or never officially ends.
How will success be measured?	Declaring victory based on meeting attendance.
What is the burden on real workers?	Improvement that improves the binder while punishing the project.

The PCC and Executive Management

Executive management must sponsor the PCC without smothering it. The executive role is to provide authority, funding, protection, and priority. It is not to define every procedure from a conference room far above the cubicles. Top-down sponsorship must enable bottom-up intelligence.

Executives often say they empower employees, but empowerment without time, authority, budget, and decision rights is decorative language. It sounds excellent in a town hall meeting and evaporates by the time the first project deadline is threatened. The PCC requires real executive support: time to meet, authority to pilot, access to stakeholders, permission to challenge broken practices, and a mechanism to escalate resistance.

The PCC and the Real Workers

The people doing the real work must see the PCC as a forum that listens rather than a court that sentences. If the PCC becomes associated only with audits, corrections, and mandatory templates, workers will naturally treat it like a process tax. But if the committee removes confusion, clarifies ownership, reduces rework, and protects teams from arbitrary change, it earns credibility.

Credibility is the true currency of process improvement. A committee can be authorized by the CIO and still be ignored by the department. A process can be posted on the portal and still be bypassed by every project. The PCC succeeds when the workers believe the process helps them deliver work with less chaos and fewer ambushes.

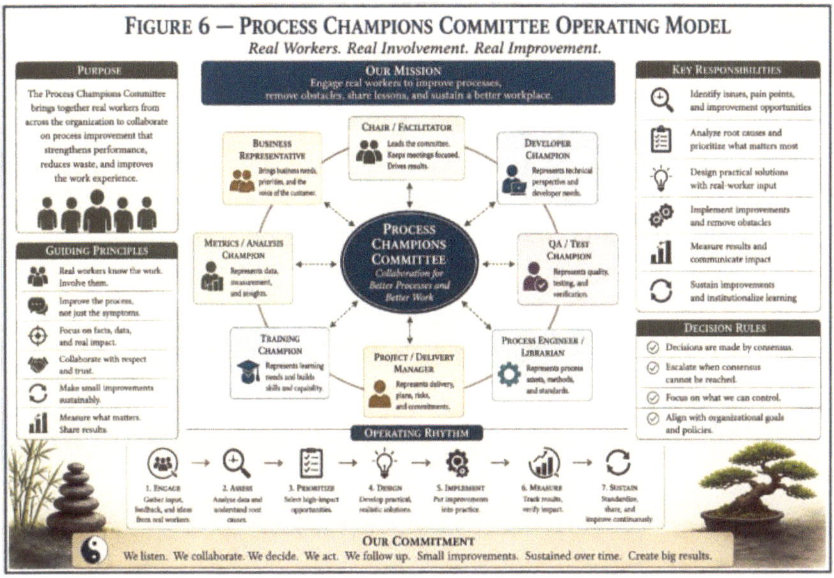

Figure 6 — Process Champions Committee Operating Model

PCC Operating Playbook

Legacy Field Observation

PCC trouble appeared when a committee existed in name but had not yet earned trust from the people doing the real work.

The practical lesson is simple: when charter combines with representation, the organization begins paying interest on process debt. That interest appears as rework, political defense, duplicate meetings, unexplained status colors, and heroic weekends that are celebrated publicly and resented privately.

IPI gives the committee a practical mandate: listen, prioritize, pilot, measure, and institutionalize only what proves useful.

Practical Guidance

- Define the normal path for charter in one page before debating the exception path for three weeks.
- Assign a named owner for representation, because anonymous ownership is where good processes go to become folklore.
- Create a visible artifact for decision log so teams can inspect the same evidence instead of arguing from memory.
- Measure issue routing only if the result will trigger a decision, correction, or meaningful conversation.
- Review executive visibility at a fixed cadence so improvement does not depend on one exhausted champion with a color-coded binder.

Legacy Insert: What Good Looks Like

A good PCC does not become a ceremonial committee. It becomes the place where real process pain is heard, prioritized, and converted into practical improvement.

Control	Focus	Minimum Evidence	Review Cadence
Control Point 1	Charter	Owner named, artifact visible, timing defined, exception path documented	Weekly during pilot; monthly after stabilization
Control Point 2	Representation	Owner named, artifact visible, timing defined, exception path documented	Weekly during pilot; monthly after stabilization
Control Point 3	Decision Log	Owner named, artifact visible, timing defined, exception path documented	Weekly during pilot; monthly after stabilization
Control Point 4	Issue Routing	Owner named, artifact visible, timing defined, exception path documented	Weekly during pilot; monthly after stabilization
Control Point 5	Executive Visibility	Owner named, artifact visible, timing defined, exception path documented	Weekly during pilot; monthly after stabilization

Mini-Case Scenario

A committee begins with confidence, agendas, volunteers, and snacks that briefly improve morale. Two months later, the team discovers that charter was assumed, representation was informal, and decision log was discussed in a hallway by people who are now on vacation. The project manager asks for accountability. The workers ask for documentation. The sponsor asks why the status was green last week. Everyone looks at the tester, because testers are apparently the archaeologists of bad decisions.

The IPI response is to give the PCC a clear charter, practical authority, and a rhythm that produces decisions instead of minutes alone.

Chapter Checklist

#	Checklist Item	Status	Notes
1	The role accountable for the activity is named.	Yes / No / Partial	Evidence or owner
2	The required artifact is defined in plain language.	Yes / No / Partial	Evidence or owner
3	The approval point is visible before downstream work begins.	Yes / No / Partial	Evidence or owner
4	The exception path is documented and time-limited.	Yes / No / Partial	Evidence or owner
5	The measurement has an action threshold.	Yes / No / Partial	Evidence or owner
6	The process can be taught to a new employee without folklore.	Yes / No / Partial	Evidence or owner

Closing Thought

The PCC is the first serious test of Intelligent Process Improvement. It reveals whether the organization truly wants process discipline or merely wants to appear disciplined. A real PCC makes process visible, practical, teachable, auditable, and adjustable. It brings the people closest to the work into the act of improving the work. In a legacy SDLC environment, that is not bureaucracy. That is harmony with a meeting agenda.

CHAPTER 6

The Bottom-Up Four-Tier Process Hierarchy

The Four-Tier Process Hierarchy is the structural backbone of IPI. It recognizes that process improvement cannot live only at the policy level, nor can it survive as scattered local habits. A mature organization needs layers. It needs executive intent, departmental standards, project procedures, and worker-level instructions that connect to one another without crushing the people underneath them.

The bottom-up principle begins with respect. It assumes that the development organization already contains much of the knowledge needed to improve itself. The people closest to the work know where the friction lives. They know which documents are useful and which documents are created for the ceremonial satisfaction of people who never read them. They know where the handoff breaks, where testing is squeezed, where requirements become fog, and where change control becomes a polite suggestion.

The purpose of the hierarchy is to capture that knowledge and arrange it into controlled process levels. Without hierarchy, process assets become a drawer full of unrelated files. With too much hierarchy, the organization becomes a museum of approvals. IPI seeks the middle: enough structure to guide work, enough simplicity to keep work moving.

IPI Reality Check

The quickest way to destroy a process hierarchy is to name every document a "standard." The second quickest way is to store all standards in a folder called "New Process Stuff FINAL FINAL Updated Use This One."

The Four Tiers

Tier 1: Corporate Process Policy

Tier	Name	Purpose	Typical Artifact
Tier 1	Corporate Process Policy	States the organization's process principles, authority, and required control points.	Corporate policy directive
Tier 2	Department Process Standard	Defines how a functional area performs its process responsibilities.	Department standard or process area guide
Tier 3	Project Procedure	Tailors the department standard to a specific project or program.	Project process plan or procedure
Tier 4	Worker Instruction and Template	Provides the practical steps, forms, checklists, and examples used by real workers.	Template, checklist, job aid, or work instruction

Tier 1 defines what the organization values and requires. It should be concise, authoritative, and stable. A corporate policy is not the place to explain every step in requirements analysis or every field in a defect report. It is the place to establish the organization's commitment to disciplined process execution.

A useful Tier 1 policy might state that all software projects must define requirements, maintain approved baselines, conduct appropriate reviews, control changes, perform independent testing, document acceptance, and retain required process artifacts. These are the governing expectations. The details belong in lower tiers.

Tier 1 must be written in a way executives can support and workers can understand. If the policy requires translation by a certification priesthood, it has already failed the IPI test.

Tier 2: Department Process Standard

Tier 2 translates corporate expectation into departmental practice. Business analysis, project management, development, testing, configuration management, release management, training, and support may each require defined standards. These standards explain the role of the function, the major activities it performs, the artifacts it produces, and the handoffs it must honor.

Tier 2 is where the legacy SDLC gains operational meaning. It prevents the familiar argument in which every function insists it did its job while the project still somehow fails. The analyst says the requirements were clear. The developer says the design changed. The tester says the build was unstable. The project manager says the status was green until it became red. Tier 2 forces each function to define what it owes the next function.

Tier 3: Project Procedure

Tier 3 recognizes that not all projects are identical. A maintenance release, a regulatory system, a commercial product, and an internal reporting tool do not require the same weight of process. However, tailoring must be controlled. Tailoring is not an escape hatch for teams that dislike documentation. Tailoring is the disciplined adjustment of process based on project risk, size, complexity, customer expectations, and contractual requirements.

The project procedure should state how the project will apply the organization's process standards. It should identify required artifacts, review points, baselines, approvals, configuration rules, testing obligations, and acceptance criteria. It should also identify justified deviations. If a project is exempt from a process requirement, the reason should be visible and approved, not whispered in the hallway.

Tier 4: Worker Instruction and Template

Tier 4 is where process becomes usable. It includes templates, checklists, examples, job aids, work instructions, naming conventions, review forms, signoff sheets, traceability matrices, defect logs, test case formats, release checklists, and other practical tools. This is the tier most directly experienced by real workers.

Tier 4 must be designed with mercy. If a template is longer than the work it supports, the template has become the project. If a checklist requires three meetings to understand, it is not a checklist. If a job aid requires a job aid, someone has gone too far.

The elegance of Tier 4 is that it makes the right behavior easier. A good template guides thinking. A good checklist prevents omission. A good work instruction reduces interpretation. These assets are not glamorous, but neither is a stable production release until one has lived through an unstable one.

IPI Reality Check

Never underestimate the spiritual power of a good checklist. It may not look impressive in an executive briefing, but it has saved more projects than the phrase "strategic alignment" ever will.

Bottom-Up Does Not Mean Leaderless

Bottom-up process improvement does not mean management disappears. It means the organization begins process definition by observing how work is actually performed, then builds guidance upward from reality. Management still approves, funds, supports, and enforces. The PCC still governs. The process librarian still controls the repository. The auditor still inspects compliance. Bottom-up is not chaos in democratic clothing. It is structured listening followed by disciplined institutionalization.

Process Asset Repository

The hierarchy requires a process asset repository. In the legacy era, this may have been a shared drive, intranet site, document management system, binder, or portal. The technology matters less than the control. There must be one official location for approved process assets, and there must be a clear distinction between draft, pilot, approved, retired, and obsolete materials.

Four-Tier Process Hierarchy in Practice

Legacy Field Observation

Hierarchy trouble appeared when workgroup practices, procedures, standards, and policy were stacked like a pyramid on paper but disconnected in daily use.

The practical lesson is simple: when policy clarity combines with procedure ownership, the organization begins paying interest on process debt. That interest appears as rework, political defense, duplicate meetings, unexplained status colors, and heroic weekends that are celebrated publicly and resented privately.

IPI connects each tier so improvement rises from real work, becomes documented guidance, then matures into shared standards and governance.

Practical Guidance

- Define the normal path for policy clarity in one page before debating the exception path for three weeks.
- Assign a named owner for procedure ownership, because anonymous ownership is where good processes go to become folklore.
- Create a visible artifact for work instruction usability so teams can inspect the same evidence instead of arguing from memory.
- Measure template control only if the result will trigger a decision, correction, or meaningful conversation.
- Review exception handling at a fixed cadence so improvement does not depend on one exhausted champion with a color-coded binder.

Legacy Insert: What Good Looks Like

A good process hierarchy lets local practice inform procedure, procedure inform standards, and standards inform governance without suffocating the work.

Control	Focus	Minimum Evidence	Review Cadence
Control Point 1	Policy Clarity	Owner named, artifact visible, timing defined, exception path documented	Weekly during pilot; monthly after stabilization
Control Point 2	Procedure Ownership	Owner named, artifact visible, timing defined, exception path documented	Weekly during pilot; monthly after stabilization
Control Point 3	Work Instruction Usability	Owner named, artifact visible, timing defined, exception path documented	Weekly during pilot; monthly after stabilization
Control Point 4	Template Control	Owner named, artifact visible, timing defined, exception path documented	Weekly during pilot; monthly after stabilization
Control Point 5	Exception Handling	Owner named, artifact visible, timing defined, exception path documented	Weekly during pilot; monthly after stabilization

Mini-Case Scenario

A workgroup begins with confidence, four tiers on a diagram, and no shared understanding of how the tiers connect. Two months later, the team discovers that policy clarity was assumed, procedure ownership was informal, and work instruction usability was discussed in a hallway by people who are now on vacation. The project manager asks for accountability. The workers ask for documentation. The sponsor asks why the status was green last week. Everyone looks at the tester, because testers are apparently the archaeologists of bad decisions.

The IPI response is to repair the connection between tiers so that standards are grounded in real work and governance supports rather than smothers.

Chapter Checklist

#	Checklist Item	Status	Notes
1	The role accountable for the activity is named.	Yes / No / Partial	Evidence or owner
2	The required artifact is defined in plain language.	Yes / No / Partial	Evidence or owner
3	The approval point is visible before downstream work begins.	Yes / No / Partial	Evidence or owner
4	The exception path is documented and time-limited.	Yes / No / Partial	Evidence or owner
5	The measurement has an action threshold.	Yes / No / Partial	Evidence or owner
6	The process can be taught to a new employee without folklore.	Yes / No / Partial	Evidence or owner

Closing Thought

The Four-Tier Process Hierarchy gives IPI its structure. It allows a legacy organization to move from vague improvement slogans to usable process architecture. It honors corporate authority, departmental expertise, project tailoring, and worker-level practicality. When the tiers align, process feels less like a burden and more like a map. When they do not align, the organization returns to the ancient tradition of arguing about which spreadsheet is the real one.

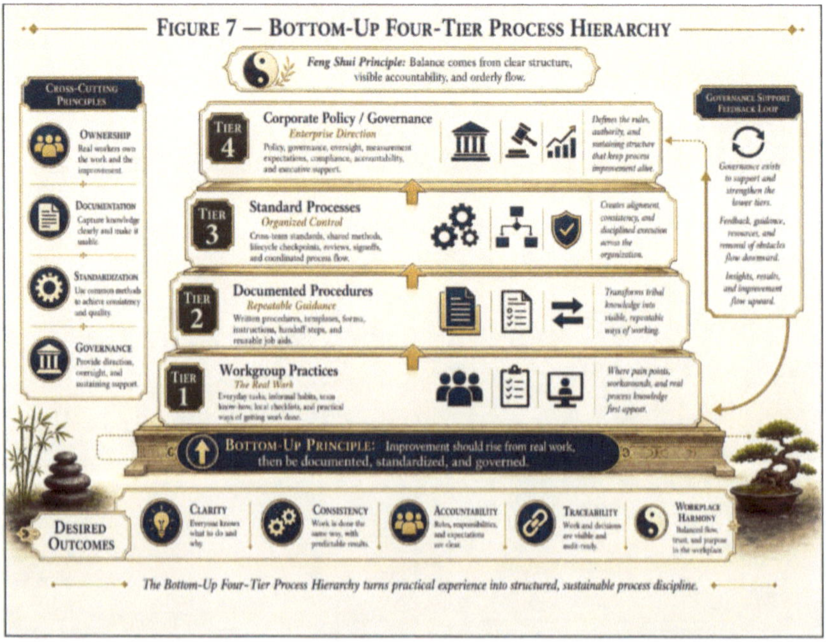

Figure 7 — Bottom-Up Four-Tier Process Hierarchy

CHAPTER 7

The IPI Process Lifecycle: Find, Fix, Measure, Maintain

The IPI Process Lifecycle reduces improvement to four practical movements: Find, Fix, Measure, and Maintain. It is intentionally simple. Legacy organizations already had enough complicated charts. What they often lacked was a disciplined way to move from problem recognition to controlled improvement without creating a process initiative so large that it required its own recovery plan.

Find means identify the real problem. Fix means define and pilot the improvement. Measure means determine whether the improvement worked. Maintain means institutionalize, train, audit, and adjust the process so the organization does not drift back into old habits.

The lifecycle is compatible with structured SDLC thinking. It has phases, artifacts, review points, roles, and controls. But unlike a massive methodology implementation, it is small enough to be understood by the people who must use it.

Find

Finding is not complaining. Complaining is easy. It requires no evidence, no root cause, and no willingness to help. Finding is the disciplined identification of a process problem that interferes with project success, product quality, worker effectiveness, customer satisfaction, or management visibility.

In the Find phase, the PCC collects issues from audits, retrospectives, project reviews, metrics, customer feedback, defect trends, schedule slippage, rework patterns, and worker observation. The goal is not to gather every grievance in the building. The goal is to identify repeatable process breakdowns that can be improved through structured intervention.

The first question is always: Is this a process problem? If a single person forgot to attach a document once, that may be a performance or training issue. If five projects cannot locate the approved requirements template, that is a process asset problem. If every release has last-minute configuration confusion, that is a process control problem. If every project starts with vague scope and heroic optimism, that is a management tradition disguised as planning.

Find Activity	Output
Collect candidate issues	Process issue log
Classify by process area	Issue category and affected roles
Identify evidence	Examples, artifacts, metrics, and observations
Determine root cause	Problem statement and causal factors
Prioritize	Severity, urgency, risk, and improvement value

IPI Reality Check

A complaint becomes a process issue when it survives evidence. “Everything is broken” is a feeling. “Three of the last four releases lacked approved rollback instructions” is a process problem with a clipboard and a pulse.

Fix

Fixing is the controlled design of improvement. This is where the PCC defines the future state, drafts revised procedures, updates templates, assigns responsibilities, and determines how the improvement will be piloted. Fixing should not begin with a mandate. It should begin with a proposed solution that can be tested.

A fix must be specific enough to change behavior. “Communicate better” is not a fix. “All requirements changes after baseline must be entered into the change request log, reviewed by the change control board, and reflected in the traceability matrix before development begins” is closer to a fix. It tells people what must happen, where it must be recorded, and when it must occur.

The best fixes reduce ambiguity. They identify who does the work, what artifact is produced, when the work begins, when it ends, who reviews it, how exceptions are handled, and where the evidence is stored. This is the grammar of legacy process engineering.

Fix Element	Question Answered
Future-state process map	How should the process flow?
Role responsibility matrix	Who performs, reviews, approves, and receives?
Artifact list	What must be created or updated?
Template or checklist	How will the worker perform the task consistently?
Pilot plan	Where will the fix be tested before institutionalization?
Training plan	How will users learn the new process?

Measure

Measuring determines whether the fix improved anything. A process that cannot be measured may still have value, but it cannot be defended. Legacy process improvement depends on evidence. The organization must be able to say whether the change reduced defects, rework, cycle time, missed handoffs, audit findings, customer complaints, or worker confusion.

Measurement should not be weaponized. Metrics exist to reveal the health of the process, not to decorate a dashboard or punish the first person whose name appears in a defect log. The Metrics Coordinator must help the PCC interpret numbers in context. A rise in reported defects may indicate worse quality, or it may indicate better testing visibility. A drop in change requests may indicate stable requirements, or it may indicate that everyone found a way to avoid the form.

Good measurement combines quantitative and qualitative evidence. Numbers tell the committee what changed. Worker feedback explains why it changed.

Maintain

Maintaining is the most neglected phase because it is less exciting than launching a new initiative. Launching feels dramatic. Maintaining feels responsible. But legacy process improvement succeeds or fails in maintenance. If the organization does not train, audit, refresh, and protect the process, the process will slowly decay into folklore.

Maintenance includes repository control, version updates, refresher training, periodic audits, metrics review, lessons learned, and retirement of obsolete assets. A process that is never retired becomes clutter. A process that is never audited becomes optional. A process that is never updated becomes historical fiction.

The Maintain phase also protects against process drift. People naturally adapt processes to project pressure. Some adaptation is healthy. Uncontrolled adaptation becomes entropy. The PCC must distinguish between intelligent tailoring and silent abandonment.

IPI Reality Check

Every process has two lives: the official life in the repository and the secret life in the cubicles. Maintain is how the PCC politely introduces them to each other before the audit does it rudely.

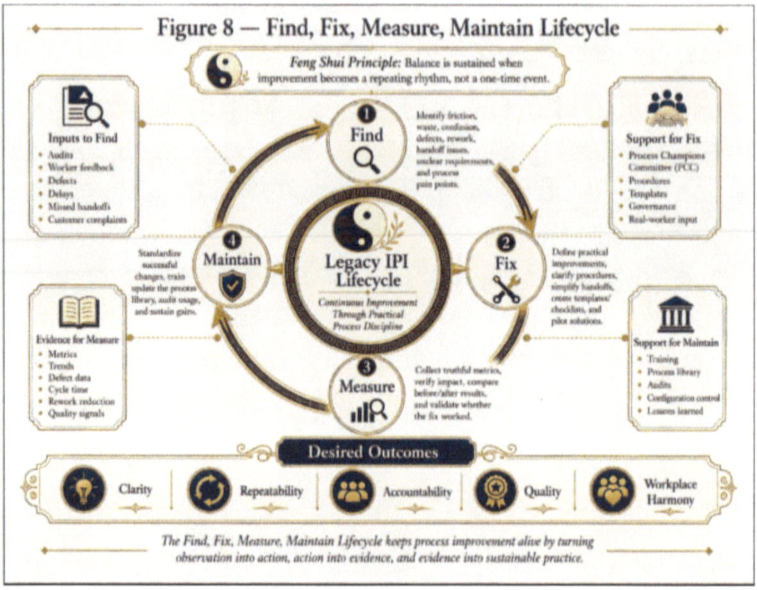

Figure 8 — Find, Fix, Measure, Maintain Lifecycle

Find, Fix, Measure, Maintain Field Guide

Legacy Field Observation

Lifecycle trouble appeared when teams were willing to find problems and fix problems, but not disciplined enough to measure and maintain the correction.

The practical lesson is simple: when root cause combines with pilot scope, the organization begins paying interest on process debt. That interest appears as rework, political defense, duplicate meetings, unexplained status colors, and heroic weekends that are celebrated publicly and resented privately.

IPI insists that improvement is a loop. A fix is not complete until it has evidence, ownership, training, and a maintenance path.

Practical Guidance

- Define the normal path for root cause in one page before debating the exception path for three weeks.
- Assign a named owner for pilot scope, because anonymous ownership is where good processes go to become folklore.
- Create a visible artifact for baseline so teams can inspect the same evidence instead of arguing from memory.
- Measure measurement only if the result will trigger a decision, correction, or meaningful conversation.
- Review sustainment at a fixed cadence so improvement does not depend on one exhausted champion with a color-coded binder.

Legacy Insert: What Good Looks Like

A good lifecycle discipline proves that finding and fixing are only half the job; measuring and maintaining are what keep the problem from returning.

Control	Focus	Minimum Evidence	Review Cadence
Control Point 1	Root Cause	Owner named, artifact visible, timing defined, exception path documented	Weekly during pilot; monthly after stabilization
Control Point 2	Pilot Scope	Owner named, artifact visible, timing defined, exception path documented	Weekly during pilot; monthly after stabilization

Control	Focus	Minimum Evidence	Review Cadence
Control Point 3	Baseline	Owner named, artifact visible, timing defined, exception path documented	Weekly during pilot; monthly after stabilization
Control Point 4	Measurement	Owner named, artifact visible, timing defined, exception path documented	Weekly during pilot; monthly after stabilization
Control Point 5	Sustainment	Owner named, artifact visible, timing defined, exception path documented	Weekly during pilot; monthly after stabilization

Mini-Case Scenario

A team begins with confidence after finding problems quickly and fixing the obvious ones. Two months later, the team discovers that root cause was assumed, pilot scope was informal, and baseline was discussed in a hallway by people who are now on vacation. The project manager asks for accountability. The workers ask for documentation. The sponsor asks why the status was green last week. Everyone looks at the tester, because testers are apparently the archaeologists of bad decisions.

The IPI response is to complete the loop: find the issue, fix the cause, measure the result, and maintain the gain until the new behavior becomes normal.

Chapter Checklist

#	Checklist Item	Status	Notes
1	The role accountable for the activity is named.	Yes / No / Partial	Evidence or owner

#	Checklist Item	Status	Notes
2	The required artifact is defined in plain language.	Yes / No / Partial	Evidence or owner
3	The approval point is visible before downstream work begins.	Yes / No / Partial	Evidence or owner
4	The exception path is documented and time-limited.	Yes / No / Partial	Evidence or owner
5	The measurement has an action threshold.	Yes / No / Partial	Evidence or owner
6	The process can be taught to a new employee without folklore.	Yes / No / Partial	Evidence or owner

Closing Thought

Find, Fix, Measure, Maintain gives IPI its motion. It prevents process improvement from becoming either endless analysis or reckless policy-making. It asks the organization to discover real problems, design practical fixes, test them, measure them, and sustain them. That may not sound glamorous, but neither does plumbing until the building floods.

CHAPTER 8

The IPI Structured SDLC Operating Model

The IPI Structured SDLC Operating Model brings the preceding chapters into the software development life cycle. The SDLC is the stage upon which process roles, artifacts, controls, and values perform. Without an SDLC, process improvement has no sequence. Without process improvement, the SDLC becomes a chart that everyone admires and no one follows.

This restored legacy chapter intentionally treats the SDLC as a phase-driven discipline. Requirements precede design. Design precedes development. Development precedes integrated testing. Testing precedes acceptance. Acceptance precedes controlled deployment. Maintenance follows release. This does not mean reality always behaves so neatly. It means the organization establishes a disciplined reference model against which reality can be managed.

The original IPI spirit recognized that no single SDLC should be worshipped as universal. However, in this legacy edition, the SDLC is still the organizing map. It is the "when" of process improvement: when artifacts are created, when reviews occur, when baselines are established, when changes are controlled, when testing begins, and when acceptance can be responsibly claimed.

The SDLC as a Process Environment

A legacy SDLC is more than a list of phases. It is a process environment. Each phase should have entry criteria, activities, roles, artifacts, reviews, measures, exit criteria, and transition rules. The transition rules matter. Many projects fail not because the phases are unknown, but because work is allowed to leave one phase before it is ready for the next.

SDLC Phase	Primary Question	Key Artifact
Initiation	Why does this project exist?	Project charter or initiation document
Requirements	What must the system do?	Requirements specification and traceability matrix
Design	How will the system satisfy the requirements?	Design specification
Development	How will the solution be built?	Code, build notes, unit test evidence
Integration and Testing	Does the solution work as intended?	Test plan, test cases, defect log, test summary
Acceptance	Will the customer accept the product?	Acceptance signoff
Deployment	How will the product be released safely?	Release package and deployment checklist
Maintenance	How will the product be sustained?	Maintenance log and change records

Requirements Phase

The requirements phase is where many projects quietly succeed or quietly fail. When requirements are vague, every later phase becomes interpretation. Design becomes guessing. Development becomes improvisation. Testing becomes archaeology. Acceptance becomes negotiation. In a legacy SDLC, requirements must be captured, reviewed, baselined, and controlled.

The Business Analyst plays a central role. The analyst must translate business need into usable requirements while managing the gap between what stakeholders say, what they mean, what they forgot, what they assume everyone knows, and what they will insist was obvious after seeing the first test build.

Requirements discipline is not glamorous. It involves patience, documentation, review, and clarification. But it prevents the ancient project curse: building exactly what was written and discovering no one wanted what was written.

Design Phase

The design phase gives shape to the solution before construction begins. In a legacy model, design is the bridge between requirements and development. It defines structure, interfaces, data, controls, screens, reports, technical constraints, and integration points. A design review should confirm that the proposed solution satisfies the approved requirements and can be built, tested, deployed, and maintained.

Development Phase

Development transforms design into working software. Even in a structured model, development requires judgment. The legacy process does not ask developers to become typists for design documents. It asks them to build within agreed boundaries, follow coding standards, perform unit testing, document significant deviations, and participate in controlled reviews.

Testing Phase

Testing is not the phase where quality is magically inserted after everyone else has finished being optimistic. Testing verifies whether the product satisfies requirements and whether the implementation is stable enough for acceptance. The testing function needs approved requirements, stable builds, test data, environments, defect management, and authority to report findings without being accused of negativity.

IPI Reality Check

A tester who finds defects is not "slowing down the project." The project was already slow; the tester merely turned on the lights.

Configuration and Change Control

Configuration management protects the integrity of the product and the process artifacts. In a legacy SDLC, baselines matter. Version control matters. Change requests matter. Release packages matter. Without configuration control, the organization cannot reliably answer basic questions: What changed? Who approved it? Which version was tested? Which build went to production? Why is the file named Final_Final_Use_This_One_v7?

Acceptance and Deployment

Acceptance confirms that the customer or authorized representative agrees the product satisfies the defined criteria. Deployment moves the product into its operational environment. These phases should not be collapsed into a frantic Friday afternoon ceremony. Acceptance requires evidence. Deployment requires planning. Rollback requires humility. Support requires preparation.

Tailoring the SDLC Without Destroying It

A structured SDLC should be tailored based on risk, complexity, regulatory needs, customer expectations, project size, and product criticality. Tailoring does not mean skipping every unpleasant activity. It means scaling the process intelligently. A small low-risk enhancement may not need the same artifact burden as a major enterprise system. But both need enough discipline to protect quality, accountability, and understanding.

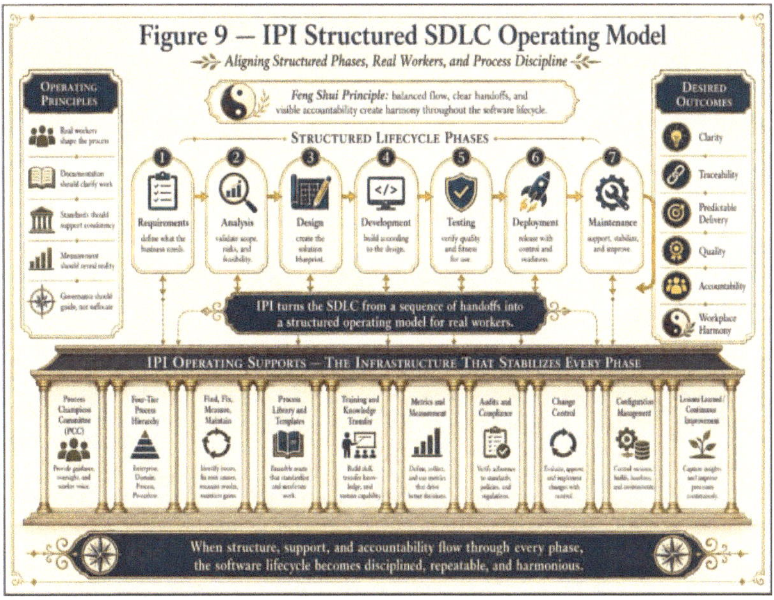

Figure 9 — IPI Structured SDLC Operating Model

Legacy SDLC Operating Model

Legacy Field Observation

Structured SDLC trouble appeared when each phase performed its own work but no one protected the flow across the entire lifecycle.

The practical lesson is simple: when requirements package combines with design review, the organization begins paying interest on process debt. That interest appears as rework, political defense, duplicate meetings, unexplained status colors, and heroic weekends that are celebrated publicly and resented privately.

IPI aligns phase activity with process supports so the lifecycle behaves like a system rather than a sequence of disconnected departments.

Practical Guidance

- Define the normal path for requirements package in one page before debating the exception path for three weeks.
- Assign a named owner for design review, because anonymous ownership is where good processes go to become folklore.
- Create a visible artifact for test planning so teams can inspect the same evidence instead of arguing from memory.
- Measure release readiness only if the result will trigger a decision, correction, or meaningful conversation.
- Review maintenance intake at a fixed cadence so improvement does not depend on one exhausted champion with a color-coded binder.

Legacy Insert: What Good Looks Like

A good structured operating model connects phases, roles, artifacts, and control points so the organization can see the whole flow.

Control	Focus	Minimum Evidence	Review Cadence
Control Point 1	Requirements Package	Owner named, artifact visible, timing defined, exception path documented	Weekly during pilot; monthly after stabilization
Control Point 2	Design Review	Owner named, artifact visible, timing defined, exception path documented	Weekly during pilot; monthly after stabilization

Control	Focus	Minimum Evidence	Review Cadence
Control Point 3	Test Planning	Owner named, artifact visible, timing defined, exception path documented	Weekly during pilot; monthly after stabilization
Control Point 4	Release Readiness	Owner named, artifact visible, timing defined, exception path documented	Weekly during pilot; monthly after stabilization
Control Point 5	Maintenance Intake	Owner named, artifact visible, timing defined, exception path documented	Weekly during pilot; monthly after stabilization

Mini-Case Scenario

A lifecycle effort begins with confidence because every phase has a box and every box has an arrow. Two months later, the team discovers that requirements package was assumed, design review was informal, and test planning was discussed in a hallway by people who are now on vacation. The project manager asks for accountability. The workers ask for documentation. The sponsor asks why the status was green last week. Everyone looks at the tester, because testers are apparently the archaeologists of bad decisions.

The IPI response is to align each SDLC phase with its supporting process infrastructure and confirm that evidence moves with the work.

Chapter Checklist

#	Checklist Item	Status	Notes
1	The role accountable for the activity is named.	Yes / No / Partial	Evidence or owner
2	The required artifact is defined	Yes / No / Partial	Evidence or owner

#	Checklist Item	Status	Notes
	in plain language.		
3	The approval point is visible before downstream work begins.	Yes / No / Partial	Evidence or owner
4	The exception path is documented and time-limited.	Yes / No / Partial	Evidence or owner
5	The measurement has an action threshold.	Yes / No / Partial	Evidence or owner
6	The process can be taught to a new employee without folklore.	Yes / No / Partial	Evidence or owner

Closing Thought

The IPI Structured SDLC Operating Model restores the dignity of sequence. It recognizes that software work becomes less chaotic when the organization knows where it is in the lifecycle, what artifact is expected, who owns the work, what review is required, and what evidence proves readiness to move forward. The SDLC is not the enemy of creativity. It is the container that keeps creativity from spilling all over production.

CHAPTER 9

Legacy Implementation Roadmap: Bringing IPI into the Organization

An IPI implementation should begin with observation, not proclamation. Too many process initiatives begin with a dramatic announcement that improvement is coming, followed by immediate suspicion from the people who have survived previous improvements. The legacy implementation roadmap is designed to earn trust by understanding the workplace before attempting to rearrange it.

The roadmap is intentionally practical. It does not require the organization to stop all work, rename every role, or purchase a new vocabulary. It asks the organization to observe the current state, form the PCC, identify priority process gaps, pilot improvements, train the workforce, measure results, and institutionalize what works.

The implementation must move fast enough to show value but slowly enough to avoid becoming another disruption. The real worker has seen enough initiatives that arrived with banners and left behind broken templates. IPI must be different. It must bring visible relief.

Phase 1: Environment Observation

The Environment Observation Phase studies how work actually happens. This includes reviewing existing artifacts, interviewing key roles, observing meetings, examining project history, looking at defects and schedule trends, and identifying informal practices that may already be effective. The goal is not to embarrass the organization. The goal is to see clearly.

Phase 2: PCC Formation

The organization forms the Process Champions Committee with a balanced mix of standing members and rotating representatives. The PCC charter is approved, meeting cadence is established, roles are assigned, and the committee receives orientation on IPI principles, the Four-Tier Process Hierarchy, and the Find, Fix, Measure, Maintain lifecycle.

Phase 3: Priority Gap Selection

The PCC selects a small number of high-value process gaps. Early wins matter. The first improvements should be visible, practical, and meaningful to the people doing the work. A good first target might be requirements baseline control, release checklist standardization, defect management cleanup, or process asset repository organization.

Phase 4: Pilot

The selected process improvement is piloted on one or more appropriate projects. The pilot should be formal enough to generate evidence but not so burdensome that it collapses under its own ceremony. The pilot team receives training, uses the new artifacts, records issues, and provides feedback to the PCC.

Phase 5: Training and Institutionalization

If the pilot succeeds, the PCC refines the process asset, obtains approval, places the official version in the repository, and trains affected roles. Training should explain not only what changed but why it changed. Adults are more likely to follow a process when they understand the problem it solves.

Phase 6: Audit and Sustainment

After adoption, the auditor and metrics coordinator assess compliance and effectiveness. The process is adjusted as needed. Sustainment prevents the organization from sliding back into its former habits the moment the next deadline appears.

Implementation Phase	Primary Output	Success Indicator
Environment Observation	Current-state findings	The organization recognizes itself in the findings.
PCC Formation	Charter and role assignments	The committee has authority and representation.
Priority Gap Selection	Improvement backlog	The first targets are specific and valuable.
Pilot	Pilot report	Workers can perform the process and provide feedback.
Training and Institutionalization	Approved assets and training record	The process is understood and accessible.
Audit and Sustainment	Compliance and effectiveness results	The process continues after launch.

IPI Reality Check

Beware the implementation plan that starts with "enterprise-wide rollout." That phrase has buried more good ideas than bad coffee and mandatory team-building combined.

The 90-Day Legacy IPI Start

A disciplined organization can begin meaningful IPI adoption within 90 days if the scope is controlled. The objective is not to fix the entire department. The objective is to establish the PCC, select priority gaps, pilot a small number of improvements, and prove that the approach can create visible order.

Timeframe	Focus	Deliverables
Days 1-15	Observation and sponsorship	Executive agreement, observation plan, initial process issue list
Days 16-30	PCC launch	Charter, membership, cadence, repository structure
Days 31-45	Gap selection	Prioritized improvement candidates and pilot selection
Days 46-65	Process definition	Draft procedure, templates, roles, metrics, training material
Days 66-80	Pilot execution	Pilot results, worker feedback, adoption issues
Days 81-90	Approval and rollout plan	Approved revisions, training plan, audit schedule

Managing Resistance

Resistance is not always rebellion. Sometimes resistance is memory. Workers remember earlier initiatives that promised improvement and delivered paperwork. They remember consultants who asked for process artifacts, disappeared into a conference room, and emerged with slides proving what everyone already knew. They remember being told to document the current process when the real problem was that no current process existed.

IPI should treat resistance as information. Ask what workers fear, what failed before, what processes already waste time, and what would make the improvement credible. Some resistance will be political. Some will be practical. Some will be justified. The PCC must listen carefully enough to tell the difference.

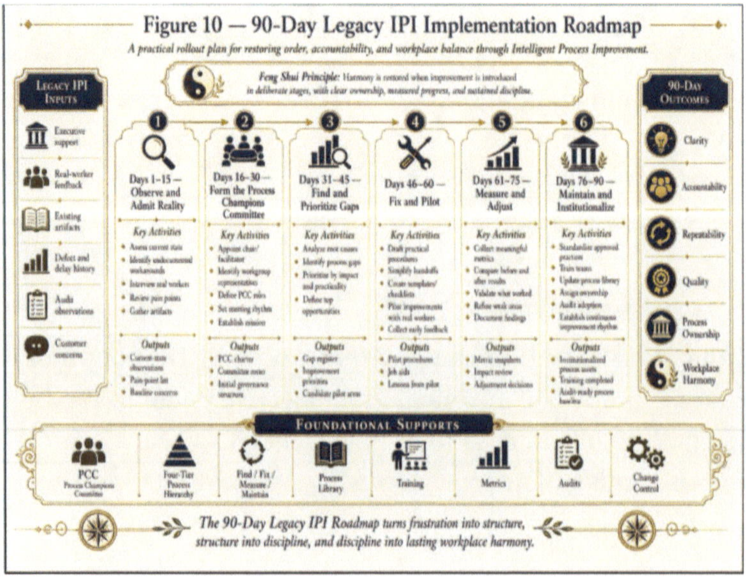

Figure 10 — 90-Day Legacy IPI Implementation Roadmap

Implementation Roadmap Deepening

Legacy Field Observation

Implementation trouble appeared when leaders wanted rapid results but had not prepared the roles, artifacts, cadence, and authority needed to sustain them.

The practical lesson is simple: when readiness combines with stakeholder map, the organization begins paying interest on process debt. That interest appears as rework, political defense, duplicate meetings, unexplained status colors, and heroic weekends that are celebrated publicly and resented privately.

IPI implementation begins small enough to be real and disciplined enough to last. The first victory must be usable, visible, and measurable.

Practical Guidance

- Define the normal path for readiness in one page before debating the exception path for three weeks.
- Assign a named owner for stakeholder map, because anonymous ownership is where good processes go to become folklore.
- Create a visible artifact for pilot selection so teams can inspect the same evidence instead of arguing from memory.
- Measure communication only if the result will trigger a decision, correction, or meaningful conversation.
- Review rollout cadence at a fixed cadence so improvement does not depend on one exhausted champion with a color-coded binder.

Legacy Insert: What Good Looks Like

A good roadmap produces early confidence without pretending that ninety days can solve ten years of process neglect.

Control	Focus	Minimum Evidence	Review Cadence
Control Point 1	Readiness	Owner named,	Weekly during

Control	Focus	Minimum Evidence	Review Cadence
		artifact visible, timing defined, exception path documented	pilot; monthly after stabilization
Control Point 2	Stakeholder Map	Owner named, artifact visible, timing defined, exception path documented	Weekly during pilot; monthly after stabilization
Control Point 3	Pilot Selection	Owner named, artifact visible, timing defined, exception path documented	Weekly during pilot; monthly after stabilization
Control Point 4	Communication	Owner named, artifact visible, timing defined, exception path documented	Weekly during pilot; monthly after stabilization
Control Point 5	Rollout Cadence	Owner named, artifact visible, timing defined, exception path documented	Weekly during pilot; monthly after stabilization

Mini-Case Scenario

A roadmap begins with confidence and a ninety-day calendar that looks peaceful from a distance. Two months later, the team discovers that readiness was assumed, stakeholder map was informal, and pilot selection was discussed in a hallway by people who are now on vacation. The project manager asks for accountability. The workers ask for documentation. The sponsor asks why the status was green last week. Everyone looks at the tester, because testers are apparently the archaeologists of bad decisions.

The IPI response is to implement in deliberate stages, protecting worker confidence while building repeatable capability.

Chapter Checklist

#	Checklist Item	Status	Notes
1	The role accountable for the activity is named.	Yes / No / Partial	Evidence or owner
2	The required artifact is defined in plain language.	Yes / No / Partial	Evidence or owner
3	The approval point is visible before downstream work begins.	Yes / No / Partial	Evidence or owner
4	The exception path is documented and time-limited.	Yes / No / Partial	Evidence or owner
5	The measurement has an action threshold.	Yes / No / Partial	Evidence or owner
6	The process can be taught to a new employee without folklore.	Yes / No / Partial	Evidence or owner

Closing Thought

Implementation is where philosophy becomes work. A legacy IPI rollout must be disciplined, visible, respectful, and evidence-driven. It must prove that process improvement can reduce frustration rather than manufacture it. The organization does not need another initiative that looks impressive at kickoff. It needs a better Monday morning.

CHAPTER 10

Process Audits, Accountability, and the Ancient Art of Proving You Actually Followed the Process

IPI Reality Check

Process improvement should lower confusion, not create a new decorative layer of confusion with a nicer logo.

Defined Process
Required Evidence
Audit Review
Finding or Confirmation
Corrective Action
Verified Improvement

Figure 10.1 - IPI Audit Accountability Flow

Why Audits Belong in a Legacy Process World

In the legacy process world, an audit is not supposed to be a corporate ambush. It is supposed to be a mirror. The problem is that many organizations hand the mirror to the wrong person, turn off the lights, invite three consultants to interpret the reflection, and then wonder why the real workers do not trust the result. An IPI audit begins with the assumption that people usually want to do good work, but the system around them may be too vague, too political, too undocumented, or too allergic to accountability to make good work repeatable.

A structured audit asks a simple question: did the organization do what it said it would do? If the answer is no, the second question is not, 'Who can we embarrass in a status meeting?' The second question is, 'Was the process usable, known, supported, measured, and realistic?' Legacy process discipline does not mean punishing the person closest to the defect. It means tracing the defect back to the missing decision, missing artifact, missing review, missing sign-off, missing environment control, or missing executive courage that allowed chaos to masquerade as speed.

The purpose of this chapter is to restore the audit to its proper place in the IPI system. It is not theater. It is not a maturity-level costume party. It is not an excuse for someone with a laminated badge to walk around saying, 'per the policy' in a tone normally reserved for airport security. It is a practical mechanism for keeping the process honest and keeping management from pretending that heroics are a strategy.

No process has ever been improved by placing an unread policy on a shared drive and announcing victory before lunch. The real work begins after the announcement, when someone asks where the template is and three people point to three different folders.

Practical Application

- Define the specific behavior in this area that the organization expects from each role before asking workers to demonstrate compliance.
- Identify the artifact or evidence that proves the behavior occurred; if no evidence exists, decide whether the process needs evidence or whether the requirement is unnecessary ceremony.
- Assign an owner who can answer questions, update guidance, and bring unresolved issues to the Process Champions Committee.
- Measure whether this discipline reduces rework, confusion, delays, defects, or repeated meetings over the next review period.

Questions for the Process Champions Committee

- Is this process step clear enough for a new worker to follow without private coaching from a veteran?

- Does the current artifact help make or prove a decision, or does it merely satisfy tradition?
- What exception would be legitimate, and who has authority to approve it?
- How will we know whether the process is improving the work environment rather than simply adding administrative weight?

The IPI Audit Principle

The IPI audit principle is direct: audit the process before blaming the people. Most workplace frustration begins when people are held accountable for processes that were never clearly defined, never communicated, never trained, never stored in a usable place, or never protected from management exceptions. If the organization cannot point to the approved process, the correct template, the official repository, and the expected evidence, then the audit finding should begin with the process system, not the worker's character.

In the legacy model, evidence matters. A project plan either exists or it does not. A requirements baseline is either approved or it is wishful thinking in Times New Roman. A change request either entered change control or it was a hallway conversation promoted to production. The audit is the calm, documented way of making those differences visible. It allows the organization to stop arguing from memory and start learning from facts.

A good IPI audit also creates protection for real workers. When a project fails because the customer changed requirements three times after sign-off, the audit should expose the change trail. When a deadline was impossible because the estimate was negotiated by optimism instead of engineering judgment, the audit should expose that. When a release failed because someone bypassed testing in the name of urgency, the audit should expose the waiver, the approver, the risk, and the business reason. Accountability without traceability is just office gossip with a spreadsheet.

No process has ever been improved by placing an unread policy on a shared drive and announcing victory before lunch. The real work begins after the announcement, when someone asks where the template is and three people point to three different folders.

Practical Application

- Define the specific behavior in this area that the organization expects from each role before asking workers to demonstrate compliance.
- Identify the artifact or evidence that proves the behavior occurred; if no evidence exists, decide whether the process needs evidence or whether the requirement is unnecessary ceremony.
- Assign an owner who can answer questions, update guidance, and bring unresolved issues to the Process Champions Committee.
- Measure whether this discipline reduces rework, confusion, delays, defects, or repeated meetings over the next review period.

Questions for the Process Champions Committee

- Is this process step clear enough for a new worker to follow without private coaching from a veteran?
- Does the current artifact help make or prove a decision, or does it merely satisfy tradition?
- What exception would be legitimate, and who has authority to approve it?
- How will we know whether the process is improving the work environment rather than simply adding administrative weight?

Audit Evidence and the Process Trail

A process trail is the sequence of evidence that shows how work moved from idea to delivery. It should be boring in the best possible way. The requirements document led to the design document. The design document led to the build plan. The build plan led to the test plan. The test results led to the release readiness review. The release approval led to deployment. The deployment led to lessons learned. When this chain exists, the organization can learn. When it does not exist, everyone attends the post-mortem with highly selective memories.

The IPI auditor does not need to become a courtroom attorney. The auditor needs to become a disciplined observer. Where is the artifact? Who approved it? Was the version controlled? Was the correct template used? Were exceptions documented? Was the customer involved at the right point? Were defects classified? Were metrics captured? Did lessons learned result in a process improvement request, or did everyone agree to 'do better next time,' which is the official anthem of unmeasured organizations?

The trail does not have to be perfect to be useful. In fact, the first audit of an informal organization often reveals that the trail looks less like a highway and more like a scavenger hunt. That is not failure. That is the baseline. The baseline is where honest improvement begins.

No process has ever been improved by placing an unread policy on a shared drive and announcing victory before lunch. The real work begins after the announcement, when someone asks where the template is and three people point to three different folders.

Practical Application

- Define the specific behavior in this area that the organization expects from each role before asking workers to demonstrate compliance.
- Identify the artifact or evidence that proves the behavior occurred; if no evidence exists, decide whether the process needs evidence or whether the requirement is unnecessary ceremony.
- Assign an owner who can answer questions, update guidance, and bring unresolved issues to the Process Champions Committee.
- Measure whether this discipline reduces rework, confusion, delays, defects, or repeated meetings over the next review period.

Questions for the Process Champions Committee

- Is this process step clear enough for a new worker to follow without private coaching from a veteran?

- Does the current artifact help make or prove a decision, or does it merely satisfy tradition?
- What exception would be legitimate, and who has authority to approve it?
- How will we know whether the process is improving the work environment rather than simply adding administrative weight?

The Auditor as Process Diplomat

The best IPI auditor is not a spy. The best auditor is a diplomat with a checklist. The role requires enough firmness to protect the process and enough emotional intelligence to avoid turning the audit into a public execution. Workers should not fear the auditor. They should fear undocumented shortcuts, invisible decisions, and emergency exceptions that become permanent operating models.

The auditor must understand the culture of the organization. Some teams hide defects because they have learned that bad news is punished. Some project managers hide schedule risk because executives prefer fictional green statuses over honest yellow ones. Some developers avoid documentation because documentation has historically been used as a weapon rather than a tool. A good auditor sees these patterns and records them as systemic issues, not merely individual failures.

In the IPI worldview, the auditor is one of the guardians of harmony. That may sound too peaceful for a role associated with findings and corrective actions, but it is exactly right. Harmony is not achieved by pretending problems do not exist. Harmony is achieved when problems can be surfaced without panic, measured without politics, and corrected without making the real workers feel like they have been drafted into a blame parade.

Practical Application

- Define the specific behavior in this area that the organization expects from each role before asking workers to demonstrate compliance.

- Identify the artifact or evidence that proves the behavior occurred; if no evidence exists, decide whether the process needs evidence or whether the requirement is unnecessary ceremony.
- Assign an owner who can answer questions, update guidance, and bring unresolved issues to the Process Champions Committee.
- Measure whether this discipline reduces rework, confusion, delays, defects, or repeated meetings over the next review period.

Questions for the Process Champions Committee

- Is this process step clear enough for a new worker to follow without private coaching from a veteran?
- Does the current artifact help make or prove a decision, or does it merely satisfy tradition?
- What exception would be legitimate, and who has authority to approve it?
- How will we know whether the process is improving the work environment rather than simply adding administrative weight?

Findings, Corrective Actions, and the Myth of the Perfect Process

An audit finding is not an insult. It is an invitation to fix something. The organization that treats findings as shameful will eventually teach everyone to hide evidence. The organization that treats findings as improvement fuel will become more disciplined without becoming more brittle. Legacy process improvement requires enough maturity to say, 'This is not working,' before the customer says it louder and with legal representation.

Each finding should be specific, evidence-based, risk-ranked, and linked to a corrective action. A vague finding such as 'requirements process needs improvement' belongs in the same drawer as inspirational posters and abandoned lunch-and-learn slides. A useful finding says that three of five sampled projects lacked approved requirements baselines before design began, resulting in rework, test delay, and customer disagreement. That is actionable. That can be fixed.

The myth of the perfect process must also be retired. A process can be disciplined without being perfect. A process can be controlled without being suffocating. A process can require evidence without demanding that every worker spend half the week feeding templates like office livestock. The IPI audit should reduce frustration, not multiply it.

Practical Application

- Define the specific behavior in this area that the organization expects from each role before asking workers to demonstrate compliance.
- Identify the artifact or evidence that proves the behavior occurred; if no evidence exists, decide whether the process needs evidence or whether the requirement is unnecessary ceremony.
- Assign an owner who can answer questions, update guidance, and bring unresolved issues to the Process Champions Committee.
- Measure whether this discipline reduces rework, confusion, delays, defects, or repeated meetings over the next review period.

Questions for the Process Champions Committee

- Is this process step clear enough for a new worker to follow without private coaching from a veteran?
- Does the current artifact help make or prove a decision, or does it merely satisfy tradition?
- What exception would be legitimate, and who has authority to approve it?
- How will we know whether the process is improving the work environment rather than simply adding administrative weight?

Audit Cadence and Governance

A legacy organization should audit on a rhythm. Monthly mini-audits keep the system aware. Quarterly process audits keep leadership honest. Release readiness audits protect production. Annual process health reviews reveal whether the system is improving or simply becoming better decorated. The cadence should be known in advance so the audit is not perceived as a surprise attack from the Department of Sudden Compliance.

Governance matters because audit findings without governance become complaints. The Process Champions Committee should own the audit calendar, review the findings, prioritize corrective actions, and report status to executive management. Executive management should not merely receive the findings; it should remove obstacles. When leadership receives the same finding three quarters in a row and does nothing, the process has not failed. Leadership has.

Audit governance also prevents overreaction. One missed template does not mean the organization needs a new methodology. One late sign-off does not mean every project manager must attend a three-day certification course in a hotel ballroom with stale muffins. The governance body separates isolated defects from systemic patterns.

Practical Application

- Define the specific behavior in this area that the organization expects from each role before asking workers to demonstrate compliance.
- Identify the artifact or evidence that proves the behavior occurred; if no evidence exists, decide whether the process needs evidence or whether the requirement is unnecessary ceremony.
- Assign an owner who can answer questions, update guidance, and bring unresolved issues to the Process Champions Committee.
- Measure whether this discipline reduces rework, confusion, delays, defects, or repeated meetings over the next review period.

Questions for the Process Champions Committee

- Is this process step clear enough for a new worker to follow without private coaching from a veteran?

- Does the current artifact help make or prove a decision, or does it merely satisfy tradition?
- What exception would be legitimate, and who has authority to approve it?
- How will we know whether the process is improving the work environment rather than simply adding administrative weight?

Implementation Summary

Area	Legacy Discipline	Evidence
Governance	Decision owner identified	PCC minutes or approval log
Artifacts	Required templates controlled	Process Library register
Training	Role impact explained	Training record or quick guide
Audit	Compliance and usefulness checked	Audit checklist and findings
Metrics	Outcome measured over time	Monthly metric trend
Improvement	Corrective action maintained	Closure record and lessons learned

IPI Reality Check

The legacy discipline works only when it helps real workers deliver quality without pretending chaos is a personality trait.

Audits and Accountability

Legacy Field Observation

Audit trouble appeared when evidence was treated as a last-minute scavenger hunt instead of a natural byproduct of disciplined work.

The practical lesson is simple: when audit plan combines with evidence list, the organization begins paying interest on process debt. That interest appears as rework, political defense, duplicate meetings, unexplained status colors, and heroic weekends that are celebrated publicly and resented privately.

IPI uses audits as mirrors. The purpose is to reveal the gap between described process and lived process, then correct the imbalance.

Practical Guidance

- Define the normal path for audit plan in one page before debating the exception path for three weeks.
- Assign a named owner for evidence list, because anonymous ownership is where good processes go to become folklore.
- Create a visible artifact for finding severity so teams can inspect the same evidence instead of arguing from memory.
- Measure corrective action only if the result will trigger a decision, correction, or meaningful conversation.
- Review closure at a fixed cadence so improvement does not depend on one exhausted champion with a color-coded binder.

Legacy Insert: What Good Looks Like

A good audit process makes evidence ordinary. The less dramatic the audit, the healthier the process.

Control	Focus	Minimum Evidence	Review Cadence
Control Point 1	Audit Plan	Owner named, artifact visible, timing defined, exception path	Weekly during pilot; monthly after stabilization

Control	Focus	Minimum Evidence	Review Cadence
		documented	
Control Point 2	Evidence List	Owner named, artifact visible, timing defined, exception path documented	Weekly during pilot; monthly after stabilization
Control Point 3	Finding Severity	Owner named, artifact visible, timing defined, exception path documented	Weekly during pilot; monthly after stabilization
Control Point 4	Corrective Action	Owner named, artifact visible, timing defined, exception path documented	Weekly during pilot; monthly after stabilization
Control Point 5	Closure	Owner named, artifact visible, timing defined, exception path documented	Weekly during pilot; monthly after stabilization

Mini-Case Scenario

An audit cycle begins with confidence until someone asks where the evidence lives. Two months later, the team discovers that audit plan was assumed, evidence list was informal, and finding severity was discussed in a hallway by people who are now on vacation. The project manager asks for accountability. The workers ask for documentation. The sponsor asks why the status was green last week. Everyone looks at the tester, because testers are apparently the archaeologists of bad decisions.

The IPI response is to treat findings as information. The question is not who failed the audit; the question is what the process is revealing.

Chapter Checklist

#	Checklist Item	Status	Notes
1	The role accountable for the activity is named.	Yes / No / Partial	Evidence or owner
2	The required artifact is defined in plain language.	Yes / No / Partial	Evidence or owner
3	The approval point is visible before downstream work begins.	Yes / No / Partial	Evidence or owner
4	The exception path is documented and time-limited.	Yes / No / Partial	Evidence or owner
5	The measurement has an action threshold.	Yes / No / Partial	Evidence or owner
6	The process can be taught to a new employee without folklore.	Yes / No / Partial	Evidence or owner

CHAPTER 11

Training, Process Memory, and the Ceremony of Teaching People What the Binder Actually Says

Approved Process
Role-Based Training
Applied Work
Audit Feedback
Updated Training
Sustained Process Memory

Figure 11.1 - Process Training Memory Loop

Training Is Not a One-Time Event

In many organizations, process training is treated like a vaccination: one injection during onboarding, followed by years of assumed immunity. The employee receives a slide deck, nods politely, forgets half of it by lunch, and is later blamed for not following a process that nobody has explained since the last reorganization. IPI rejects this ritual. Training is not an event. Training is how the organization preserves process memory.

Process memory is the organization's ability to remember how it does work even after people leave, managers rotate, tools change, projects end, and the expert who knew everything moves to another department and suddenly becomes unreachable except through vague calendar invites. A mature legacy process organization does not rely entirely on tribal knowledge. It captures the steps, roles, artifacts, decision points, and lessons in a way that can be taught, practiced, and improved.

This is especially important in structured software development because the process crosses many hands. Business analysts, project managers, developers, testers, configuration managers, release managers, trainers, customers, and executives all touch the lifecycle. If only one group understands the process, the process will fail at the handoff. Training turns individual understanding into shared operating discipline.

Practical Application

- Define the specific behavior in this area that the organization expects from each role before asking workers to demonstrate compliance.
- Identify the artifact or evidence that proves the behavior occurred; if no evidence exists, decide whether the process needs evidence or whether the requirement is unnecessary ceremony.
- Assign an owner who can answer questions, update guidance, and bring unresolved issues to the Process Champions Committee.
- Measure whether this discipline reduces rework, confusion, delays, defects, or repeated meetings over the next review period.

Questions for the Process Champions Committee

- Is this process step clear enough for a new worker to follow without private coaching from a veteran?
- Does the current artifact help make or prove a decision, or does it merely satisfy tradition?
- What exception would be legitimate, and who has authority to approve it?
- How will we know whether the process is improving the work environment rather than simply adding administrative weight?

The Process Trainer Role

The Process Trainer is more than a person who reads slides out loud while the room quietly checks email. The Process Trainer is the translator between the official process and the working reality of the teams. This role takes the approved process and makes it understandable to the people who must use it. The trainer explains not only what must be done, but why it matters, where the artifacts live, who approves them, and how the work protects the project from future confusion.

In the IPI model, the trainer is tied to reputation. If people do not understand the process, the organization will eventually produce inconsistent work. Inconsistent work becomes inconsistent quality. Inconsistent quality becomes damaged reputation. A company can spend a fortune on branding, but if releases repeatedly fail because nobody understood the approval gate, the market will not be impressed by the logo.

Training must also be role-specific. Executives do not need the same training as testers. Testers do not need the same training as customers. Customers do not need to know how the configuration baseline is stored, but they do need to know what it means when they sign off on requirements. The trainer designs learning paths that respect the different responsibilities within the lifecycle.

Practical Application

- Define the specific behavior in this area that the organization expects from each role before asking workers to demonstrate compliance.
- Identify the artifact or evidence that proves the behavior occurred; if no evidence exists, decide whether the process needs evidence or whether the requirement is unnecessary ceremony.
- Assign an owner who can answer questions, update guidance, and bring unresolved issues to the Process Champions Committee.
- Measure whether this discipline reduces rework, confusion, delays, defects, or repeated meetings over the next review period.

Questions for the Process Champions Committee

- Is this process step clear enough for a new worker to follow without private coaching from a veteran?
- Does the current artifact help make or prove a decision, or does it merely satisfy tradition?
- What exception would be legitimate, and who has authority to approve it?
- How will we know whether the process is improving the work environment rather than simply adding administrative weight?

Learning Paths for the Legacy SDLC

A complete training system should contain learning paths for each major process role. The business analyst learns requirements elicitation, documentation, traceability, baseline management, and change impact analysis. The project manager learns planning, risk tracking, status reporting, phase gate preparation, and escalation. The developer learns coding standards, build discipline, peer review expectations, configuration management handoffs, and defect response. The tester learns test planning, test case design, defect classification, regression control, and release evidence. The executive learns governance, process waiver responsibility, funding discipline, and the dangerous difference between urgency and panic.

The organization should also train people on handoffs. Many process failures happen not inside a role, but between roles. The requirements handoff to design. The design handoff to development. The development handoff to testing. The testing handoff to release. The release handoff to operations or support. Each handoff should have entry criteria, exit criteria, artifacts, owners, and acceptance expectations. A handoff without expectations is a polite way of throwing work over a wall and hoping the person on the other side has a helmet.

Legacy training does not need to be dull. It can include scenarios, sample artifacts, before-and-after examples, process walk-throughs, and role-play exercises. A worker learns more from one realistic project story than from twenty abstract slides about compliance. The trainer should use examples from the organization's own history, with names changed to protect the guilty, the innocent, and the consultant who billed forty hours to rename a folder.

Practical Application

- Define the specific behavior in this area that the organization expects from each role before asking workers to demonstrate compliance.
- Identify the artifact or evidence that proves the behavior occurred; if no evidence exists, decide whether the process needs evidence or whether the requirement is unnecessary ceremony.
- Assign an owner who can answer questions, update guidance, and bring unresolved issues to the Process Champions Committee.
- Measure whether this discipline reduces rework, confusion, delays, defects, or repeated meetings over the next review period.

Questions for the Process Champions Committee

- Is this process step clear enough for a new worker to follow without private coaching from a veteran?
- Does the current artifact help make or prove a decision, or does it merely satisfy tradition?
- What exception would be legitimate, and who has authority to approve it?
- How will we know whether the process is improving the work environment rather than simply adding administrative weight?

The Training Repository

Every training system needs a repository. The repository should contain approved process documents, templates, examples, quick reference guides, recorded sessions, frequently asked questions, waiver forms, audit criteria, and lessons learned. It should be organized so a normal human being can find something without needing a treasure map, a committee vote, and three passwords.

The Process Librarian and the Process Trainer should work as partners. The trainer teaches the material; the librarian protects the authoritative source. If the trainer teaches one version and the repository stores another, confusion multiplies. If the repository becomes a dumping ground for every draft ever created, workers will stop trusting it. The repository should make the correct process easier to follow than the incorrect process.

Good repositories are curated. Obsolete templates are retired. Current templates are clearly labeled. Examples are marked as examples. Policies are separated from guidance. Process changes are dated and communicated. A repository without curation is not a library; it is an attic. And no one should have to climb through an attic to find the test plan template before a release readiness meeting.

Practical Application

- Define the specific behavior in this area that the organization expects from each role before asking workers to demonstrate compliance.
- Identify the artifact or evidence that proves the behavior occurred; if no evidence exists, decide whether the process needs evidence or whether the requirement is unnecessary ceremony.
- Assign an owner who can answer questions, update guidance, and bring unresolved issues to the Process Champions Committee.
- Measure whether this discipline reduces rework, confusion, delays, defects, or repeated meetings over the next review period.

Questions for the Process Champions Committee

- Is this process step clear enough for a new worker to follow without private coaching from a veteran?
- Does the current artifact help make or prove a decision, or does it merely satisfy tradition?
- What exception would be legitimate, and who has authority to approve it?

- How will we know whether the process is improving the work environment rather than simply adding administrative weight?

Certification Without the Cult Behavior

Internal certification can be useful when it proves practical readiness. It becomes dangerous when it turns into a miniature religion. IPI supports the idea that people should demonstrate understanding before serving in process roles, especially rotating Process Champions Committee roles. However, the goal is competence, not credential theater. A worker should not need a special handshake, a framed certificate, and three initials after their name to know where the change request form is located.

A simple internal certification can include a short knowledge check, a sample artifact review, and a scenario-based exercise. For example, a new PCC representative might be asked to identify missing evidence in a project file, classify a process gap, and recommend a corrective action. That is useful. Asking them to memorize obscure terminology so they can pass a test and forget it immediately is less useful.

Training should empower workers. It should not create a new class system in which process priests speak in acronyms and ordinary workers wait for interpretation. IPI was designed to make process improvement understandable, not to build another velvet rope around common sense.

Practical Application

- Define the specific behavior in this area that the organization expects from each role before asking workers to demonstrate compliance.
- Identify the artifact or evidence that proves the behavior occurred; if no evidence exists, decide whether the process needs evidence or whether the requirement is unnecessary ceremony.
- Assign an owner who can answer questions, update guidance, and bring unresolved issues to the Process Champions Committee.

- Measure whether this discipline reduces rework, confusion, delays, defects, or repeated meetings over the next review period.

Questions for the Process Champions Committee

- Is this process step clear enough for a new worker to follow without private coaching from a veteran?
- Does the current artifact help make or prove a decision, or does it merely satisfy tradition?
- What exception would be legitimate, and who has authority to approve it?
- How will we know whether the process is improving the work environment rather than simply adding administrative weight?

Measuring Training Effectiveness

Training should be measured. Attendance is not enough. Attendance only proves that a person was physically or digitally present while information occurred nearby. Effectiveness requires evidence that the learner can apply the process. Measures may include quiz results, artifact quality improvement, reduction in repeated questions, fewer audit findings related to misunderstanding, faster onboarding, and improved phase gate readiness.

The Metrics Coordinator should help define training measures that matter. If defects are repeatedly traced back to poor requirements, training should focus on requirements quality and traceability. If releases fail because deployment evidence is incomplete, training should focus on release readiness. Training should be tied to actual organizational pain, not to whatever topic has the nicest slide template.

The best sign of training effectiveness is not that workers can recite the process. It is that the process becomes visible in their work. The artifacts improve. The handoffs become cleaner. The questions become sharper. The meetings become shorter because the evidence is already prepared. That is when training stops being an obligation and becomes a form of workplace relief.

Practical Application

- Define the specific behavior in this area that the organization expects from each role before asking workers to demonstrate compliance.
- Identify the artifact or evidence that proves the behavior occurred; if no evidence exists, decide whether the process needs evidence or whether the requirement is unnecessary ceremony.
- Assign an owner who can answer questions, update guidance, and bring unresolved issues to the Process Champions Committee.
- Measure whether this discipline reduces rework, confusion, delays, defects, or repeated meetings over the next review period.

Questions for the Process Champions Committee

- Is this process step clear enough for a new worker to follow without private coaching from a veteran?
- Does the current artifact help make or prove a decision, or does it merely satisfy tradition?
- What exception would be legitimate, and who has authority to approve it?
- How will we know whether the process is improving the work environment rather than simply adding administrative weight?

IPI Reality Check

The legacy discipline works only when it helps real workers deliver quality without pretending chaos is a personality trait.

Training and Process Memory

Legacy Field Observation

Training trouble appeared when the process had been announced but not transferred into the knowledge of the people expected to follow it.

The practical lesson is simple: when training matrix combines with role-based learning, the organization begins paying interest on process debt. That interest appears as rework, political defense, duplicate meetings, unexplained status colors, and heroic weekends that are celebrated publicly and resented privately.

IPI treats training as process memory transfer. Knowledge must move from binders and veterans into repeatable worker capability.

Practical Guidance

- Define the normal path for training matrix in one page before debating the exception path for three weeks.
- Assign a named owner for role-based learning, because anonymous ownership is where good processes go to become folklore.
- Create a visible artifact for orientation so teams can inspect the same evidence instead of arguing from memory.
- Measure knowledge retention only if the result will trigger a decision, correction, or meaningful conversation.
- Review effectiveness checks at a fixed cadence so improvement does not depend on one exhausted champion with a color-coded binder.

Legacy Insert: What Good Looks Like

A good training system teaches the reason for the process as clearly as it teaches the steps.

Control	Focus	Minimum Evidence	Review Cadence
Control Point 1	Training Matrix	Owner named, artifact visible, timing defined, exception path documented	Weekly during pilot; monthly after stabilization
Control Point 2	Role-Based Learning	Owner named, artifact visible, timing defined, exception path	Weekly during pilot; monthly after stabilization

Control	Focus	Minimum Evidence	Review Cadence
		documented	
Control Point 3	Orientation	Owner named, artifact visible, timing defined, exception path documented	Weekly during pilot; monthly after stabilization
Control Point 4	Knowledge Retention	Owner named, artifact visible, timing defined, exception path documented	Weekly during pilot; monthly after stabilization
Control Point 5	Effectiveness Checks	Owner named, artifact visible, timing defined, exception path documented	Weekly during pilot; monthly after stabilization

Mini-Case Scenario

A training rollout begins with confidence because invitations were sent and slides were attached. Two months later, the team discovers that training matrix was assumed, role-based learning was informal, and orientation was discussed in a hallway by people who are now on vacation. The project manager asks for accountability. The workers ask for documentation. The sponsor asks why the status was green last week. Everyone looks at the tester, because testers are apparently the archaeologists of bad decisions.

The IPI response is to teach the process in usable pieces and verify that people can perform it without private folklore.

Chapter Checklist

#	Checklist Item	Status	Notes
1	The role accountable for the activity is named.	Yes / No / Partial	Evidence or owner
2	The required artifact is defined in plain language.	Yes / No / Partial	Evidence or owner
3	The approval point is visible before downstream work begins.	Yes / No / Partial	Evidence or owner
4	The exception path is documented and time-limited.	Yes / No / Partial	Evidence or owner
5	The measurement has an action threshold.	Yes / No / Partial	Evidence or owner
6	The process can be taught to a new employee without folklore.	Yes / No / Partial	Evidence or owner

CHAPTER 12

The Process Library, Artifact Control, and the Sacred War Against Folder Chaos

Policy
Procedure
Template
Checklist
Example
Archive

Figure 12.1 - Process Library Control Model

The Process Library as Organizational Memory

The Process Library is where the organization's disciplined memory lives. It is the official home of process descriptions, templates, policies, forms, checklists, examples, phase gate criteria, audit standards, and lessons learned. Without it, the organization becomes dependent on hallway conversations, personal folders, email archaeology, and the one person who knows where the 'real' template is stored. That person will eventually go on vacation during a production emergency.

In the IPI structure, the Process Librarian protects partnership because the library allows many roles to collaborate from a shared source of truth. The business analyst, tester, project manager, auditor, trainer, and configuration manager must not each maintain separate versions of reality. When teams use different templates, different definitions, and different approval expectations, the project does not gain flexibility. It gains confusion with attachments.

The Process Library should feel practical, not ceremonial. It should help people do the work. If workers avoid the library because it is too complicated, outdated, or bloated with irrelevant material, the library has failed. A library that nobody uses is not a process asset. It is a museum of good intentions.

Practical Application

- Define the specific behavior in this area that the organization expects from each role before asking workers to demonstrate compliance.
- Identify the artifact or evidence that proves the behavior occurred; if no evidence exists, decide whether the process needs evidence or whether the requirement is unnecessary ceremony.
- Assign an owner who can answer questions, update guidance, and bring unresolved issues to the Process Champions Committee.
- Measure whether this discipline reduces rework, confusion, delays, defects, or repeated meetings over the next review period.

Questions for the Process Champions Committee

- Is this process step clear enough for a new worker to follow without private coaching from a veteran?
- Does the current artifact help make or prove a decision, or does it merely satisfy tradition?
- What exception would be legitimate, and who has authority to approve it?
- How will we know whether the process is improving the work environment rather than simply adding administrative weight?

Artifact Classes and Ownership

Every artifact in the process system should have an owner. Ownership means someone is responsible for maintaining the artifact, reviewing it, retiring it when obsolete, and answering questions about its intended use. Without ownership, artifacts become orphans. Orphan artifacts wander through projects causing confusion, especially when one template says 'Approved By' and another says 'Sign-Off Authority' and a third one is a spreadsheet from 2004 with a macro that only works on one laptop.

Artifacts should be classified. Policies define mandatory rules. Procedures define repeatable steps. Templates standardize output. Checklists guide verification. Examples demonstrate completed work. Job aids help workers remember common tasks. Metrics reports show performance. Waivers document exceptions. Lessons learned capture improvement opportunities. Mixing all of these together without labels forces workers to interpret the process each time they use it.

The Process Librarian should maintain an artifact register. The register lists the artifact name, purpose, owner, version, approval date, related process, required roles, storage location, and retirement status. This may sound painfully structured until the day someone asks which release checklist was valid during the failed deployment. On that day, the artifact register becomes a beautiful thing.

Practical Application

- Define the specific behavior in this area that the organization expects from each role before asking workers to demonstrate compliance.
- Identify the artifact or evidence that proves the behavior occurred; if no evidence exists, decide whether the process needs evidence or whether the requirement is unnecessary ceremony.
- Assign an owner who can answer questions, update guidance, and bring unresolved issues to the Process Champions Committee.
- Measure whether this discipline reduces rework, confusion, delays, defects, or repeated meetings over the next review period.

Questions for the Process Champions Committee

- Is this process step clear enough for a new worker to follow without private coaching from a veteran?

- Does the current artifact help make or prove a decision, or does it merely satisfy tradition?
- What exception would be legitimate, and who has authority to approve it?
- How will we know whether the process is improving the work environment rather than simply adding administrative weight?

Version Control for Process Assets

Process assets require version control. A process document is not an inspirational essay. It governs work. If the organization cannot tell which version was active at the time of a project decision, auditability collapses. A legacy process organization should know when a template changed, who approved the change, what changed, why it changed, and whether teams were notified.

Version control does not have to be complicated. Each artifact should have a version number, date, owner, approval authority, change summary, and effective date. Drafts should be visibly marked as drafts. Retired versions should be archived but not used. Current versions should be easy to find. The most recent approved template should not be hidden underneath four folders named 'New,' 'New Final,' 'Final Final,' and 'Use This One Maybe.'

The Configuration Manager and Process Librarian share an important relationship here. Configuration management protects product and project baselines; process library control protects the baseline of how the organization works. Both disciplines prevent uncontrolled change from pretending to be progress.

Practical Application

- Define the specific behavior in this area that the organization expects from each role before asking workers to demonstrate compliance.
- Identify the artifact or evidence that proves the behavior occurred; if no evidence exists, decide whether the process needs evidence or whether the requirement is unnecessary ceremony.

- Assign an owner who can answer questions, update guidance, and bring unresolved issues to the Process Champions Committee.
- Measure whether this discipline reduces rework, confusion, delays, defects, or repeated meetings over the next review period.

Questions for the Process Champions Committee

- Is this process step clear enough for a new worker to follow without private coaching from a veteran?
- Does the current artifact help make or prove a decision, or does it merely satisfy tradition?
- What exception would be legitimate, and who has authority to approve it?
- How will we know whether the process is improving the work environment rather than simply adding administrative weight?

The Folder Structure Problem

Every organization eventually develops folder chaos. It starts innocently: one shared drive, one project folder, one helpful person creating a folder named 'Misc.' From there the entropy accelerates. Soon there are folders named 'Old,' 'Older,' 'Archive,' 'Archive2,' 'Do Not Use,' 'Use This,' 'Real Use This,' and 'Backup From Steve Before He Left.' Somewhere inside this maze is the approved requirements template, probably hiding next to a birthday flyer and a PDF no one can open.

IPI treats folder structure as a process issue. If the storage system is confusing, workers will create shortcuts. If workers create shortcuts, artifacts scatter. If artifacts scatter, audits become painful. If audits become painful, management assumes workers are noncompliant. In reality, the organization may have built a document maze and then punished people for not escaping quickly enough.

The Process Library should use plain categories aligned to the lifecycle: governance, requirements, project planning, design, development, testing, configuration management, release management, training, audits, metrics, waivers, and lessons learned. Within each category, documents should be sorted by type and status. The goal is boring clarity. Boring clarity is underrated until the night before a gate review.

Practical Application

- Define the specific behavior in this area that the organization expects from each role before asking workers to demonstrate compliance.
- Identify the artifact or evidence that proves the behavior occurred; if no evidence exists, decide whether the process needs evidence or whether the requirement is unnecessary ceremony.
- Assign an owner who can answer questions, update guidance, and bring unresolved issues to the Process Champions Committee.
- Measure whether this discipline reduces rework, confusion, delays, defects, or repeated meetings over the next review period.

Questions for the Process Champions Committee

- Is this process step clear enough for a new worker to follow without private coaching from a veteran?
- Does the current artifact help make or prove a decision, or does it merely satisfy tradition?
- What exception would be legitimate, and who has authority to approve it?
- How will we know whether the process is improving the work environment rather than simply adding administrative weight?

Artifact Quality Standards

A template is not valuable merely because it exists. It must produce useful information. Many organizations confuse document quantity with process maturity. They create templates so large that filling them out becomes a separate project. Workers respond by pasting generic language into every section, which produces the comforting illusion of completeness while communicating almost nothing.

IPI artifact quality standards should test whether an artifact is clear, necessary, usable, owned, reviewed, and connected to a decision. If a document does not support a decision, handoff, baseline, approval, audit, or learning need, the organization should question why it exists. Every required artifact should earn its place in the lifecycle. Legacy discipline does not mean paperwork worship. It means evidence with purpose.

A strong artifact tells the next person what they need to know to continue the work. A weak artifact says, 'See attached,' and the attachment says, 'TBD.' The Process Librarian cannot fix every weak artifact alone, but the library can define quality expectations so teams understand what good looks like.

Practical Application

- Define the specific behavior in this area that the organization expects from each role before asking workers to demonstrate compliance.
- Identify the artifact or evidence that proves the behavior occurred; if no evidence exists, decide whether the process needs evidence or whether the requirement is unnecessary ceremony.
- Assign an owner who can answer questions, update guidance, and bring unresolved issues to the Process Champions Committee.
- Measure whether this discipline reduces rework, confusion, delays, defects, or repeated meetings over the next review period.

Questions for the Process Champions Committee

- Is this process step clear enough for a new worker to follow without private coaching from a veteran?
- Does the current artifact help make or prove a decision, or does it merely satisfy tradition?

- What exception would be legitimate, and who has authority to approve it?
- How will we know whether the process is improving the work environment rather than simply adding administrative weight?

Retiring Artifacts with Dignity

Old artifacts should be retired with dignity and firmness. Nothing creates confusion like obsolete documents that remain available in active folders. Workers will use the wrong file because it is familiar, because it is shorter, because it has their favorite formatting, or because someone emailed it to them in 2011 and said, 'This is the one we use.' The library must prevent nostalgia from becoming noncompliance.

Retirement does not mean deletion without trace. Retired artifacts should be archived with dates and reasons. The organization may need historical evidence for audits or lessons learned. But retired materials should not compete with current materials. They belong in the archive, not in the active process path.

The Process Librarian's quiet power is the power to reduce confusion. In a workplace filled with meetings, emails, status decks, and urgent requests, clarity is a form of mercy. The library should be where workers go to feel less lost, not more lost.

Practical Application

- Define the specific behavior in this area that the organization expects from each role before asking workers to demonstrate compliance.
- Identify the artifact or evidence that proves the behavior occurred; if no evidence exists, decide whether the process needs evidence or whether the requirement is unnecessary ceremony.
- Assign an owner who can answer questions, update guidance, and bring unresolved issues to the Process Champions Committee.
- Measure whether this discipline reduces rework, confusion, delays, defects, or repeated meetings over the next review period.

Questions for the Process Champions Committee

- Is this process step clear enough for a new worker to follow without private coaching from a veteran?
- Does the current artifact help make or prove a decision, or does it merely satisfy tradition?
- What exception would be legitimate, and who has authority to approve it?
- How will we know whether the process is improving the work environment rather than simply adding administrative weight?

Implementation Summary

Area	Legacy Discipline	Evidence
Governance	Decision owner identified	PCC minutes or approval log
Artifacts	Required templates controlled	Process Library register
Training	Role impact explained	Training record or quick guide
Audit	Compliance and usefulness checked	Audit checklist and findings
Metrics	Outcome measured over time	Monthly metric trend
Improvement	Corrective action maintained	Closure record and lessons learned

Process Library and Artifact Control

Legacy Field Observation

Process-library trouble appeared when the organization had many documents but no confidence that the right one could be found, trusted, and used.

The practical lesson is simple: when repository map combines with version history, the organization begins paying interest on process debt. That interest appears as rework, political defense, duplicate meetings, unexplained status colors, and heroic weekends that are celebrated publicly and resented privately.

IPI treats the library as the house of process knowledge. If the house is cluttered, people stop visiting.

Practical Guidance

- Define the normal path for repository map in one page before debating the exception path for three weeks.
- Assign a named owner for version history, because anonymous ownership is where good processes go to become folklore.
- Create a visible artifact for owner so teams can inspect the same evidence instead of arguing from memory.
- Measure approval date only if the result will trigger a decision, correction, or meaningful conversation.
- Review retirement at a fixed cadence so improvement does not depend on one exhausted champion with a color-coded binder.

Legacy Insert: What Good Looks Like

A good process library makes the approved way of working easier to locate than the unofficial workaround.

Control	Focus	Minimum Evidence	Review Cadence
Control Point 1	Repository Map	Owner named, artifact visible, timing defined, exception path documented	Weekly during pilot; monthly after stabilization
Control Point 2	Version History	Owner named, artifact visible, timing defined, exception path documented	Weekly during pilot; monthly after stabilization
Control Point 3	Owner	Owner named, artifact visible, timing defined, exception path documented	Weekly during pilot; monthly after stabilization

Control	Focus	Minimum Evidence	Review Cadence
Control Point 4	Approval Date	Owner named, artifact visible, timing defined, exception path documented	Weekly during pilot; monthly after stabilization
Control Point 5	Retirement	Owner named, artifact visible, timing defined, exception path documented	Weekly during pilot; monthly after stabilization

Mini-Case Scenario

A library cleanup begins with confidence and a folder structure that seems logical to the person who created it. Two months later, the team discovers that repository map was assumed, version history was informal, and owner was discussed in a hallway by people who are now on vacation. The project manager asks for accountability. The workers ask for documentation. The sponsor asks why the status was green last week. Everyone looks at the tester, because testers are apparently the archaeologists of bad decisions.

The IPI response is to clean the repository, assign ownership, retire obsolete assets, and make the approved path visible.

Chapter Checklist

#	Checklist Item	Status	Notes
1	The role accountable for the activity is named.	Yes / No / Partial	Evidence or owner
2	The required artifact is defined in plain language.	Yes / No / Partial	Evidence or owner
3	The approval point is visible before	Yes / No / Partial	Evidence or owner

#	Checklist Item	Status	Notes
	downstream work begins.		
4	The exception path is documented and time-limited.	Yes / No / Partial	Evidence or owner
5	The measurement has an action threshold.	Yes / No / Partial	Evidence or owner
6	The process can be taught to a new employee without folklore.	Yes / No / Partial	Evidence or owner

CHAPTER 13

Metrics, Measurement, and the Dangerous Difference Between Data and Decorative Numbers

IPI Reality Check

Process improvement should lower confusion, not create a new decorative layer of confusion with a nicer logo.

Visible Pain
Baseline Measure
Trend Analysis
Decision
Corrective Action
Sustained Improvement

Figure 13.1 - Metrics from Pain to Improvement

Why Measurement Matters

Measurement is where process improvement either becomes real or becomes office decoration. A process improvement initiative without measures can still produce meetings, templates, committees, posters, slogans, and a very impressive kickoff deck. What it cannot produce is proof. IPI requires measurement because the organization must know whether the process is reducing frustration, improving quality, stabilizing delivery, and making work more predictable.

Legacy process discipline respects numbers, but it should not worship them blindly. A metric is only useful when it is connected to a decision. Defect counts matter if they reveal quality trends. Cycle time matters if it helps identify bottlenecks. Rework matters if it exposes weak requirements or design instability. Audit findings matter if they lead to corrective actions. A number that does not change behavior is just a decorative number wearing a tie.

The Metrics Coordinator plays a central role in keeping measurement healthy. This role gathers, analyzes, maintains, and communicates process metrics. The goal is not to embarrass teams. The goal is to make the health of the process visible. If the process is sick, the organization should know before the customer becomes the diagnostic instrument.

Practical Application

- Define the specific behavior in this area that the organization expects from each role before asking workers to demonstrate compliance.
- Identify the artifact or evidence that proves the behavior occurred; if no evidence exists, decide whether the process needs evidence or whether the requirement is unnecessary ceremony.
- Assign an owner who can answer questions, update guidance, and bring unresolved issues to the Process Champions Committee.
- Measure whether this discipline reduces rework, confusion, delays, defects, or repeated meetings over the next review period.

Questions for the Process Champions Committee

- Is this process step clear enough for a new worker to follow without private coaching from a veteran?
- Does the current artifact help make or prove a decision, or does it merely satisfy tradition?
- What exception would be legitimate, and who has authority to approve it?
- How will we know whether the process is improving the work environment rather than simply adding administrative weight?

The Measurement Trap

The measurement trap occurs when organizations begin measuring what is easy instead of what is meaningful. Counting documents is easy. Knowing whether the documents improved decisions is harder. Counting defects is easy. Understanding why defects escaped is harder. Counting meetings is easy. Admitting that half of them exist because the process is unclear is harder. IPI measurement must resist the temptation to collect numbers simply because numbers look mature.

Another trap is using metrics as a weapon. When metrics are used primarily to punish, workers learn to manage the numbers instead of improving the work. Defects get reclassified. Estimates get padded. Risks get hidden until they become issues. Status reports become fiction. The organization may appear healthier on paper while becoming more fragile in reality. That is not measurement. That is theater with decimals.

The third trap is measuring too much too soon. A new process improvement initiative does not need a dashboard with forty-seven indicators and three colors for each executive mood. It needs a small set of baseline measures tied to the most painful problems. Start with the pain. Measure what would prove improvement. Expand only when the organization can actually use the information.

Practical Application

- Define the specific behavior in this area that the organization expects from each role before asking workers to demonstrate compliance.
- Identify the artifact or evidence that proves the behavior occurred; if no evidence exists, decide whether the process needs evidence or whether the requirement is unnecessary ceremony.
- Assign an owner who can answer questions, update guidance, and bring unresolved issues to the Process Champions Committee.
- Measure whether this discipline reduces rework, confusion, delays, defects, or repeated meetings over the next review period.

Questions for the Process Champions Committee

- Is this process step clear enough for a new worker to follow without private coaching from a veteran?
- Does the current artifact help make or prove a decision, or does it merely satisfy tradition?
- What exception would be legitimate, and who has authority to approve it?
- How will we know whether the process is improving the work environment rather than simply adding administrative weight?

Core IPI Metrics

Core IPI metrics should connect to process stability, quality, accountability, and worker frustration. Process stability may include schedule variance, phase gate readiness, change request volume, and baseline volatility. Quality may include defect density, escaped defects, rework percentage, test coverage, and defect aging. Accountability may include approval timeliness, waiver frequency, audit finding closure, and artifact completeness. Worker frustration may include repeated blockers, unclear ownership, excessive rework, and time spent searching for information.

Some leaders resist measuring frustration because it sounds soft. That is usually because they have never calculated the cost of turnover, rework, missed deadlines, bad morale, and meetings created to discuss problems that should have been prevented by process clarity. Frustration is not merely emotional noise. It is often the leading indicator of process failure. By the time the defect report arrives, the workers usually knew something was wrong weeks earlier.

A balanced metric set should include leading and lagging indicators. Leading indicators warn before failure: unclear requirements, late approvals, missing test data, unstable environments, open risks, unresolved dependencies. Lagging indicators confirm after failure: missed dates, production defects, customer complaints, emergency releases. Mature organizations watch both. Immature organizations ignore leading indicators and then act surprised when lagging indicators arrive wearing boots.

Practical Application

- Define the specific behavior in this area that the organization expects from each role before asking workers to demonstrate compliance.
- Identify the artifact or evidence that proves the behavior occurred; if no evidence exists, decide whether the process needs evidence or whether the requirement is unnecessary ceremony.
- Assign an owner who can answer questions, update guidance, and bring unresolved issues to the Process Champions Committee.
- Measure whether this discipline reduces rework, confusion, delays, defects, or repeated meetings over the next review period.

Questions for the Process Champions Committee

- Is this process step clear enough for a new worker to follow without private coaching from a veteran?
- Does the current artifact help make or prove a decision, or does it merely satisfy tradition?
- What exception would be legitimate, and who has authority to approve it?
- How will we know whether the process is improving the work environment rather than simply adding administrative weight?

The Metrics Review Meeting

The metrics review meeting should be disciplined, short, and decision-oriented. It is not a place to admire charts. It is a place to interpret evidence and choose action. The meeting should answer four questions: What changed? Why did it change? What risk does it create? What action will we take? If the meeting cannot answer those questions, it may be a slide show pretending to be governance.

The Metrics Coordinator should prepare the data, but process owners should explain the meaning. A spike in change requests may mean requirements were unstable, or it may mean the change process finally became visible. A rise in audit findings may mean compliance worsened, or it may mean auditors are finally looking in the right places. Numbers require context. Without context, metrics become a rumor in numeric form.

Executives should attend metrics reviews prepared to remove obstacles, not merely request better colors on the dashboard. If the same bottleneck appears month after month, the issue may require funding, staffing, priority decisions, or policy clarification. Metrics without executive action can create more cynicism than no metrics at all. Workers quickly notice when the dashboard is used to monitor them but not to support them.

Practical Application

- Define the specific behavior in this area that the organization expects from each role before asking workers to demonstrate compliance.
- Identify the artifact or evidence that proves the behavior occurred; if no evidence exists, decide whether the process needs evidence or whether the requirement is unnecessary ceremony.
- Assign an owner who can answer questions, update guidance, and bring unresolved issues to the Process Champions Committee.
- Measure whether this discipline reduces rework, confusion, delays, defects, or repeated meetings over the next review period.

Questions for the Process Champions Committee

- Is this process step clear enough for a new worker to follow without private coaching from a veteran?
- Does the current artifact help make or prove a decision, or does it merely satisfy tradition?

- What exception would be legitimate, and who has authority to approve it?
- How will we know whether the process is improving the work environment rather than simply adding administrative weight?

Measurement and Feng Shui Values

The original IPI concept linked process roles to values such as unity, balance, prosperity, reputation, partnership, creativity, progress, promotion, longevity, health, and wholistic support. Measurement can support these values when designed properly. Unity can be measured by handoff success and cross-team issue resolution. Balance can be measured by workload stability and change control discipline. Reputation can be measured by defect escape rates and customer acceptance. Longevity can be measured by release stability and maintainability. Health can be measured by process sustainability and worker friction.

This may sound unusual to readers who expect process books to discuss only documents and gates. But the values matter because they remind the organization that process improvement is not just about producing artifacts. It is about creating a more harmonious way of working. The metrics should reveal whether the process is supporting people or merely producing more evidence of their exhaustion.

A good metric set turns values into observable behavior. If partnership matters, measure whether groups collaborate at defined handoffs. If creativity matters, measure whether improvement suggestions are captured and implemented. If progress matters, measure whether gaps are closing over time. If accountability matters, measure whether decisions are documented and owned. Values without measures become slogans. Measures without values become surveillance. IPI needs both.

Practical Application

- Define the specific behavior in this area that the organization expects from each role before asking workers to demonstrate compliance.

- Identify the artifact or evidence that proves the behavior occurred; if no evidence exists, decide whether the process needs evidence or whether the requirement is unnecessary ceremony.
- Assign an owner who can answer questions, update guidance, and bring unresolved issues to the Process Champions Committee.
- Measure whether this discipline reduces rework, confusion, delays, defects, or repeated meetings over the next review period.

Questions for the Process Champions Committee

- Is this process step clear enough for a new worker to follow without private coaching from a veteran?
- Does the current artifact help make or prove a decision, or does it merely satisfy tradition?
- What exception would be legitimate, and who has authority to approve it?
- How will we know whether the process is improving the work environment rather than simply adding administrative weight?

Building the Legacy Dashboard

A legacy dashboard should be clear enough to print, discuss, and archive. It should not require an interactive interface, a data engineer, and a minor miracle to understand. The dashboard should show the few measures that matter most: requirements stability, open risks, audit findings, defect trends, change requests, phase gate readiness, release readiness, and corrective action closure.

Each measure should have a definition, source, owner, frequency, threshold, and interpretation guidance. Without definitions, teams will debate the metric instead of the work. What counts as rework? What counts as an escaped defect? When is a requirement considered approved? Who determines whether a corrective action is closed? These questions must be answered before the numbers appear in front of executives.

The dashboard should also include narrative. A chart shows what happened. A short explanation tells why it matters. The best metrics reports combine numbers with plain-language interpretation. This protects leadership from staring at a downward trend line and asking whether down is good. In process improvement, clarity is always cheaper than confusion.

Practical Application

- Define the specific behavior in this area that the organization expects from each role before asking workers to demonstrate compliance.
- Identify the artifact or evidence that proves the behavior occurred; if no evidence exists, decide whether the process needs evidence or whether the requirement is unnecessary ceremony.
- Assign an owner who can answer questions, update guidance, and bring unresolved issues to the Process Champions Committee.
- Measure whether this discipline reduces rework, confusion, delays, defects, or repeated meetings over the next review period.

Questions for the Process Champions Committee

- Is this process step clear enough for a new worker to follow without private coaching from a veteran?
- Does the current artifact help make or prove a decision, or does it merely satisfy tradition?
- What exception would be legitimate, and who has authority to approve it?
- How will we know whether the process is improving the work environment rather than simply adding administrative weight?

Implementation Summary

Area	Legacy Discipline	Evidence
Governance	Decision owner identified	PCC minutes or approval log
Artifacts	Required templates controlled	Process Library register
Training	Role impact explained	Training record or quick guide
Audit	Compliance and usefulness checked	Audit checklist and findings
Metrics	Outcome measured over time	Monthly metric trend
Improvement	Corrective action maintained	Closure record and lessons learned

IPI Reality Check

The legacy discipline works only when it helps real workers deliver quality without pretending chaos is a personality trait.

Metrics and Measurement

Legacy Field Observation

Metrics trouble appeared when the numbers were collected with great care but did not change a single decision, behavior, or process weakness.

The practical lesson is simple: when definition combines with source, the organization begins paying interest on process debt. That interest appears as rework, political defense, duplicate meetings, unexplained status colors, and heroic weekends that are celebrated publicly and resented privately.

IPI asks every metric to earn its place by informing a decision, revealing a trend, or guiding improvement.

Practical Guidance

- Define the normal path for definition in one page before debating the exception path for three weeks.
- Assign a named owner for source, because anonymous ownership is where good processes go to become folklore.
- Create a visible artifact for frequency so teams can inspect the same evidence instead of arguing from memory.
- Measure interpretation only if the result will trigger a decision, correction, or meaningful conversation.
- Review action threshold at a fixed cadence so improvement does not depend on one exhausted champion with a color-coded binder.

Legacy Insert: What Good Looks Like

A good measurement system tells the truth in time for the organization to act.

Control	Focus	Minimum Evidence	Review Cadence
Control Point 1	Definition	Owner named, artifact visible, timing defined, exception path documented	Weekly during pilot; monthly after stabilization
Control Point 2	Source	Owner named, artifact visible, timing defined, exception path documented	Weekly during pilot; monthly after stabilization
Control Point 3	Frequency	Owner named, artifact visible, timing defined, exception path documented	Weekly during pilot; monthly after stabilization
Control Point 4	Interpretation	Owner named, artifact visible, timing defined, exception path documented	Weekly during pilot; monthly after stabilization
Control Point 5	Action Threshold	Owner named, artifact visible, timing defined, exception path documented	Weekly during pilot; monthly after stabilization

Mini-Case Scenario

A metrics effort begins with confidence, charts, and a dashboard that looks executive-ready. Two months later, the team discovers that definition was assumed, source was informal, and frequency was discussed in a hallway by people who are now on vacation. The project manager asks for accountability. The workers ask for documentation. The sponsor asks why the status was green last week. Everyone looks at the tester, because testers are apparently the archaeologists of bad decisions.

The IPI response is to define what the metric means, what action it triggers, and who is accountable for responding.

Chapter Checklist

#	Checklist Item	Status	Notes
1	The role accountable for the activity is named.	Yes / No / Partial	Evidence or owner
2	The required artifact is defined in plain language.	Yes / No / Partial	Evidence or owner
3	The approval point is visible before downstream work begins.	Yes / No / Partial	Evidence or owner
4	The exception path is documented and time-limited.	Yes / No / Partial	Evidence or owner
5	The measurement has an action threshold.	Yes / No / Partial	Evidence or owner
6	The process can be taught to a new employee without folklore.	Yes / No / Partial	Evidence or owner

CHAPTER 14

Sustaining IPI: Change Control, Waivers, Lessons Learned, and the Discipline of Not Sliding Backward

Change Control
Waivers
Lessons Learned
Audit Follow-Up
Training Refresh
Annual Health Review

Figure 14.1 - IPI Sustainability Cycle

The Sustainability Problem

Many process improvement initiatives begin with energy and end with a folder. The kickoff is enthusiastic. The posters are printed. The committee is formed. The first templates are created. Then real work returns, deadlines tighten, leaders rotate, budgets shift, and the process quietly becomes optional. Eventually a new executive arrives and announces a new initiative to solve the same problems the last initiative was supposed to solve. The organization applauds politely because it has seen this movie before.

Sustaining IPI requires more than initial implementation. It requires change control, governance, training refresh, audit follow-up, metrics review, waiver discipline, and lessons learned. Sustainability is the art of keeping the process alive after the excitement fades. It is less glamorous than launching a new program, but it is far more valuable.

The legacy process mindset understands that discipline must be maintained. A process is not mature because it was written. It is mature because it survives pressure. It still works when schedules are tight. It still works when customers are demanding. It still works when executives are impatient. It still works when the project manager says, 'Can we skip this just this once?' and everyone knows what the waiver process requires.

Practical Application

- Define the specific behavior in this area that the organization expects from each role before asking workers to demonstrate compliance.
- Identify the artifact or evidence that proves the behavior occurred; if no evidence exists, decide whether the process needs evidence or whether the requirement is unnecessary ceremony.
- Assign an owner who can answer questions, update guidance, and bring unresolved issues to the Process Champions Committee.
- Measure whether this discipline reduces rework, confusion, delays, defects, or repeated meetings over the next review period.

Questions for the Process Champions Committee

- Is this process step clear enough for a new worker to follow without private coaching from a veteran?
- Does the current artifact help make or prove a decision, or does it merely satisfy tradition?
- What exception would be legitimate, and who has authority to approve it?
- How will we know whether the process is improving the work environment rather than simply adding administrative weight?

Process Change Control

Process change control protects the process from casual alteration. Without it, every project modifies the process to suit its immediate preferences. Over time, the organization no longer has a process; it has local customs. One team uses version 2 of the requirements template.

Another team uses an Excel checklist. A third team uses a whiteboard photograph. A fourth team claims to be exempt because their project is 'different,' which is the traditional mating call of uncontrolled variation.

A process change request should describe the current process, the proposed change, the reason, the affected roles, the affected artifacts, the risk, the training impact, the repository impact, the audit impact, and the recommended effective date. The Process Champions Committee reviews the request and decides whether to approve, reject, defer, or pilot the change.

Piloting is important. Not every process change should be rolled out across the organization immediately. A pilot allows the organization to test whether the change improves the work or simply sounds impressive in a meeting. The pilot should have success criteria, time limits, feedback collection, and a decision point. Otherwise the pilot becomes a permanent exception with better branding.

Practical Application

- Define the specific behavior in this area that the organization expects from each role before asking workers to demonstrate compliance.
- Identify the artifact or evidence that proves the behavior occurred; if no evidence exists, decide whether the process needs evidence or whether the requirement is unnecessary ceremony.
- Assign an owner who can answer questions, update guidance, and bring unresolved issues to the Process Champions Committee.
- Measure whether this discipline reduces rework, confusion, delays, defects, or repeated meetings over the next review period.

Questions for the Process Champions Committee

- Is this process step clear enough for a new worker to follow without private coaching from a veteran?
- Does the current artifact help make or prove a decision, or does it merely satisfy tradition?

- What exception would be legitimate, and who has authority to approve it?
- How will we know whether the process is improving the work environment rather than simply adding administrative weight?

The Executive Process Waiver

The Executive Process Waiver exists because reality occasionally refuses to respect the process calendar. There are legitimate emergencies, contractual obligations, customer demands, legal issues, and business risks that may require deviation from the standard process. IPI does not pretend otherwise. A mature process system allows exceptions. It simply refuses to let exceptions become invisible.

A waiver should document what is being waived, why it is being waived, who approved it, what risk is being accepted, how the risk will be mitigated, when the waiver expires, and what follow-up action is required. The waiver should be visible to the Process Champions Committee and included in audit review. If leadership wants to bypass a gate, leadership should own the decision. That is not bureaucracy. That is accountability.

The waiver is also a cultural test. If waivers are rare, documented, and reviewed, the process is likely healthy. If waivers are frequent, vague, and treated as normal, the organization has a process on paper and chaos in practice. A process that can be waived every time it becomes inconvenient is not a process. It is a suggestion with a header.

Practical Application

- Define the specific behavior in this area that the organization expects from each role before asking workers to demonstrate compliance.
- Identify the artifact or evidence that proves the behavior occurred; if no evidence exists, decide whether the process needs evidence or whether the requirement is unnecessary ceremony.
- Assign an owner who can answer questions, update guidance, and bring unresolved issues to the Process Champions Committee.

- Measure whether this discipline reduces rework, confusion, delays, defects, or repeated meetings over the next review period.

Questions for the Process Champions Committee

- Is this process step clear enough for a new worker to follow without private coaching from a veteran?
- Does the current artifact help make or prove a decision, or does it merely satisfy tradition?
- What exception would be legitimate, and who has authority to approve it?
- How will we know whether the process is improving the work environment rather than simply adding administrative weight?

Lessons Learned Without the Group Therapy Spiral

Lessons learned are essential, but they are often mishandled. In some organizations, the lessons learned meeting becomes group therapy with action items nobody owns. In others, it becomes a blame session disguised as reflection. In still others, it becomes a ceremonial closing document created after everyone has mentally moved on to the next emergency. IPI lessons learned must be practical, evidence-based, and connected to process improvement.

A useful lessons learned session asks what went well, what did not go well, what surprised the team, what evidence supports the conclusion, what process gap was revealed, what corrective action is recommended, who owns the action, and when it will be reviewed. The output should feed the Process Champions Committee. Lessons that do not feed improvement are simply stories told near the end of a project.

The organization should also distinguish between project lessons and process lessons. A project lesson may be specific to a customer, vendor, tool, or team. A process lesson reveals a repeatable gap or improvement opportunity across projects. Both matter, but process lessons should become candidates for process change, training updates, template refinement, or audit focus.

Practical Application

- Define the specific behavior in this area that the organization expects from each role before asking workers to demonstrate compliance.
- Identify the artifact or evidence that proves the behavior occurred; if no evidence exists, decide whether the process needs evidence or whether the requirement is unnecessary ceremony.
- Assign an owner who can answer questions, update guidance, and bring unresolved issues to the Process Champions Committee.
- Measure whether this discipline reduces rework, confusion, delays, defects, or repeated meetings over the next review period.

Questions for the Process Champions Committee

- Is this process step clear enough for a new worker to follow without private coaching from a veteran?
- Does the current artifact help make or prove a decision, or does it merely satisfy tradition?
- What exception would be legitimate, and who has authority to approve it?
- How will we know whether the process is improving the work environment rather than simply adding administrative weight?

Preventing Process Decay

Process decay happens slowly. At first, one artifact is skipped because the team is busy. Then one gate review is shortened because everyone is in a hurry. Then one waiver is verbal because the executive is traveling. Then one template is replaced because the old one is easier. Six months later the organization wonders why quality slipped. Process decay rarely announces itself. It simply removes one discipline at a time until the structure becomes decorative.

The Process Champions Committee should monitor signs of decay. Are audit findings repeating? Are waivers increasing? Are templates being bypassed? Are lessons learned ignored? Are metrics no longer reviewed? Are new employees confused? Are old informal habits returning? These signs should trigger corrective action before the process collapses into another annual initiative.

Sustainability also requires leadership attention. Leaders do not need to micromanage every artifact, but they must reinforce the importance of process discipline. When executives bypass the process casually, workers learn that the process is optional. When executives support the process under pressure, workers learn that discipline is real. The culture follows what leadership tolerates, not what the policy says.

Practical Application

- Define the specific behavior in this area that the organization expects from each role before asking workers to demonstrate compliance.
- Identify the artifact or evidence that proves the behavior occurred; if no evidence exists, decide whether the process needs evidence or whether the requirement is unnecessary ceremony.
- Assign an owner who can answer questions, update guidance, and bring unresolved issues to the Process Champions Committee.
- Measure whether this discipline reduces rework, confusion, delays, defects, or repeated meetings over the next review period.

Questions for the Process Champions Committee

- Is this process step clear enough for a new worker to follow without private coaching from a veteran?
- Does the current artifact help make or prove a decision, or does it merely satisfy tradition?
- What exception would be legitimate, and who has authority to approve it?

- How will we know whether the process is improving the work environment rather than simply adding administrative weight?

The Annual IPI Health Review

An annual IPI health review gives the organization a structured opportunity to step back. The review should examine audit results, metric trends, training effectiveness, process change history, waiver patterns, repository health, stakeholder feedback, and major lessons learned. It should identify what to keep, what to improve, what to retire, and what to simplify.

The annual review should not become a festival of PowerPoint. It should produce decisions. Which processes are working? Which templates are burdensome? Which metrics are useful? Which roles need clearer responsibilities? Which parts of the lifecycle are most fragile? Which improvement priorities matter most for the next year?

Sustained IPI is not static. It is controlled evolution. The organization improves the process without abandoning the legacy discipline that gives the process its strength. It adapts while still respecting baselines, gates, evidence, training, and accountability. That is the difference between living process and fashionable chaos.

Practical Application

- Define the specific behavior in this area that the organization expects from each role before asking workers to demonstrate compliance.
- Identify the artifact or evidence that proves the behavior occurred; if no evidence exists, decide whether the process needs evidence or whether the requirement is unnecessary ceremony.
- Assign an owner who can answer questions, update guidance, and bring unresolved issues to the Process Champions Committee.
- Measure whether this discipline reduces rework, confusion, delays, defects, or repeated meetings over the next review period.

Questions for the Process Champions Committee

- Is this process step clear enough for a new worker to follow without private coaching from a veteran?
- Does the current artifact help make or prove a decision, or does it merely satisfy tradition?
- What exception would be legitimate, and who has authority to approve it?
- How will we know whether the process is improving the work environment rather than simply adding administrative weight?

Implementation Summary

Area	Legacy Discipline	Evidence
Governance	Decision owner identified	PCC minutes or approval log
Artifacts	Required templates controlled	Process Library register
Training	Role impact explained	Training record or quick guide
Audit	Compliance and usefulness checked	Audit checklist and findings
Metrics	Outcome measured over time	Monthly metric trend
Improvement	Corrective action maintained	Closure record and lessons learned

Change Control and Waivers

Legacy Field Observation

Sustainability trouble appeared when an improvement worked once but was never embedded into ownership, training, measurement, or review.

The practical lesson is simple: when change board combines with impact analysis, the organization begins paying interest on process debt. That interest appears as rework, political defense, duplicate meetings, unexplained status colors, and heroic weekends that are celebrated publicly and resented privately.

IPI keeps gains from sliding backward by attaching improvement to ownership, cadence, and evidence.

Practical Guidance

- Define the normal path for change board in one page before debating the exception path for three weeks.
- Assign a named owner for impact analysis, because anonymous ownership is where good processes go to become folklore.

- Create a visible artifact for waiver expiry so teams can inspect the same evidence instead of arguing from memory.
- Measure lessons learned only if the result will trigger a decision, correction, or meaningful conversation.
- Review retrofit at a fixed cadence so improvement does not depend on one exhausted champion with a color-coded binder.

Legacy Insert: What Good Looks Like

A good sustainability system keeps yesterday's improvement from becoming tomorrow's forgotten initiative.

Control	Focus	Minimum Evidence	Review Cadence
Control Point 1	Change Board	Owner named, artifact visible, timing defined, exception path documented	Weekly during pilot; monthly after stabilization
Control Point 2	Impact Analysis	Owner named, artifact visible, timing defined, exception path documented	Weekly during pilot; monthly after stabilization
Control Point 3	Waiver Expiry	Owner named, artifact visible, timing defined, exception path documented	Weekly during pilot; monthly after stabilization
Control Point 4	Lessons Learned	Owner named, artifact visible, timing defined, exception path documented	Weekly during pilot; monthly after stabilization
Control Point 5	Retrofit	Owner named, artifact visible, timing defined, exception path documented	Weekly during pilot; monthly after stabilization

Mini-Case Scenario

A sustainability effort begins with confidence because the pilot worked once and everyone applauded. Two months later, the team discovers that change board was assumed, impact analysis was informal, and waiver expiry was discussed in a hallway by people who are now on vacation. The project manager asks for accountability. The workers ask for documentation. The sponsor asks why the status was green last week. Everyone looks at the tester, because testers are apparently the archaeologists of bad decisions.

The IPI response is to embed the improvement into review cadence, training, audit evidence, and change control.

Chapter Checklist

#	Checklist Item	Status	Notes
1	The role accountable for the activity is named.	Yes / No / Partial	Evidence or owner
2	The required artifact is defined in plain language.	Yes / No / Partial	Evidence or owner
3	The approval point is visible before downstream work begins.	Yes / No / Partial	Evidence or owner
4	The exception path is documented and time-limited.	Yes / No / Partial	Evidence or owner
5	The measurement has an action threshold.	Yes / No / Partial	Evidence or owner
6	The process can be taught to a new employee without folklore.	Yes / No / Partial	Evidence or owner

CHAPTER 15

Final Words, Legacy Principles, and the Workplace Improvement Process for Real Workers

Real Workers
Clear Process
Useful Evidence
Measured Outcomes
Balanced Accountability
Harmony

Figure 15.1 - Workplace Improvement Process

Why the Legacy Still Matters

The legacy still matters because structure still matters. Even in a world that changes language every few years, software work still depends on understanding needs, planning work, controlling change, building carefully, testing honestly, releasing responsibly, and learning from outcomes. The names may change, the tools may change, the meeting titles may change, but the underlying need for disciplined work does not disappear.

This restored legacy manuscript does not argue that every old practice was perfect. It argues that the best parts of structured software development deserve preservation. Requirements should be understood before they are built. Designs should be reviewed before they are implemented. Changes should be assessed before they are accepted. Testing should produce evidence before release. Production should not be treated like a casino. Workers should not be blamed for chaos created by unclear management decisions.

The original IPI spirit remains powerful because it was written from the viewpoint of real workers. It did not worship methodology. It did not pretend that process improvement happens because executives announce it. It understood cubicle frustration, deadline panic, consultant theater, documentation overload, political status reports, and the quiet exhaustion of people trying to deliver quality in an environment that rewards urgency over clarity. That honesty is why the legacy remains valuable.

Practical Application

- Define the specific behavior in this area that the organization expects from each role before asking workers to demonstrate compliance.
- Identify the artifact or evidence that proves the behavior occurred; if no evidence exists, decide whether the process needs evidence or whether the requirement is unnecessary ceremony.
- Assign an owner who can answer questions, update guidance, and bring unresolved issues to the Process Champions Committee.
- Measure whether this discipline reduces rework, confusion, delays, defects, or repeated meetings over the next review period.

Questions for the Process Champions Committee

- Is this process step clear enough for a new worker to follow without private coaching from a veteran?
- Does the current artifact help make or prove a decision, or does it merely satisfy tradition?
- What exception would be legitimate, and who has authority to approve it?
- How will we know whether the process is improving the work environment rather than simply adding administrative weight?

The Real Worker Principle

The Real Worker Principle is simple: the people who do the work must be central to improving the work. Process improvement fails when it is done to workers rather than with workers. The most beautiful process map in the world will fail if it ignores the daily reality of the people who must follow it. The most expensive consultant deck in the world cannot replace the knowledge of the analyst, tester, developer, release manager, trainer, and support person who can tell you exactly where the process breaks at 4:45 p.m. on a Friday.

Real workers are not anti-process. They are anti-nonsense. They resist processes that slow them down without improving quality. They resist templates that ask for information nobody uses. They resist meetings that exist because the previous meeting failed to make a decision. They resist leadership mandates that create accountability downward but not upward. Give real workers a process that solves actual pain, and many of them will become its strongest defenders.

IPI honors real workers by giving them a place in the system. The Process Champions Committee includes rotating representatives. The lifecycle includes feedback. The audit looks at the process before blaming the person. The metrics reveal systemic friction. The library makes the right work easier to find. The training explains why the process matters. This is how process improvement becomes workplace improvement.

Practical Application

- Define the specific behavior in this area that the organization expects from each role before asking workers to demonstrate compliance.
- Identify the artifact or evidence that proves the behavior occurred; if no evidence exists, decide whether the process needs evidence or whether the requirement is unnecessary ceremony.
- Assign an owner who can answer questions, update guidance, and bring unresolved issues to the Process Champions Committee.
- Measure whether this discipline reduces rework, confusion, delays, defects, or repeated meetings over the next review period.

Questions for the Process Champions Committee

- Is this process step clear enough for a new worker to follow without private coaching from a veteran?
- Does the current artifact help make or prove a decision, or does it merely satisfy tradition?
- What exception would be legitimate, and who has authority to approve it?
- How will we know whether the process is improving the work environment rather than simply adding administrative weight?

The Workplace Improvement Process

If IPI had another name, it could be Workplace Improvement Process. That name captures the deeper purpose. Software process improvement is not merely about the software. It is about improving the environment in which software is created. The work environment shapes the work product. A chaotic environment produces chaotic outcomes. A disciplined, balanced, respectful environment has a better chance of producing stable systems and sane workers.

A Workplace Improvement Process asks whether the organization has created the conditions for good work. Are priorities clear? Are requirements stable enough to build? Are changes controlled? Are roles understood? Are artifacts useful? Are handoffs clean? Are metrics honest? Are executives accountable? Are customers engaged? Are workers trained? Are lessons learned applied? These questions matter as much as any methodology label.

The Feng Shui metaphor remains useful because it reminds the organization to look at flow, balance, energy, and environment. In software terms, flow means work can move through the lifecycle without constant obstruction. Balance means accountability exists without micromanagement. Energy means the work environment does not drain every ounce of motivation from the people doing the work. Environment means the organization is designed to support quality rather than merely demand it.

Practical Application

- Define the specific behavior in this area that the organization expects from each role before asking workers to demonstrate compliance.
- Identify the artifact or evidence that proves the behavior occurred; if no evidence exists, decide whether the process needs evidence or whether the requirement is unnecessary ceremony.
- Assign an owner who can answer questions, update guidance, and bring unresolved issues to the Process Champions Committee.
- Measure whether this discipline reduces rework, confusion, delays, defects, or repeated meetings over the next review period.

Questions for the Process Champions Committee

- Is this process step clear enough for a new worker to follow without private coaching from a veteran?
- Does the current artifact help make or prove a decision, or does it merely satisfy tradition?
- What exception would be legitimate, and who has authority to approve it?
- How will we know whether the process is improving the work environment rather than simply adding administrative weight?

The Legacy Principles Summary

The restored legacy manuscript can be summarized through a set of principles. First, process exists to help people produce repeatable quality, not to impress auditors. Second, accountability must be visible at every level, including executive decisions. Third, documentation should support decisions, handoffs, baselines, and learning. Fourth, change must be controlled because uncontrolled change is one of the fastest ways to convert a project plan into a historical artifact.

Fifth, metrics must be meaningful and used for improvement, not intimidation. Sixth, audits should reveal process health, not merely hunt for worker mistakes. Seventh, training must preserve process memory. Eighth, the process library must serve as the trusted source of truth. Ninth, waivers must be documented because exceptions are business decisions, not invisible shortcuts. Tenth, lessons learned must feed process improvement or they are just project folklore.

These principles are not fashionable. That is their strength. They do not depend on buzzwords. They depend on discipline, clarity, and common sense. In the legacy world, common sense was often buried under methodology language, but it was always there waiting to be rediscovered.

Practical Application

- Define the specific behavior in this area that the organization expects from each role before asking workers to demonstrate compliance.
- Identify the artifact or evidence that proves the behavior occurred; if no evidence exists, decide whether the process needs evidence or whether the requirement is unnecessary ceremony.
- Assign an owner who can answer questions, update guidance, and bring unresolved issues to the Process Champions Committee.
- Measure whether this discipline reduces rework, confusion, delays, defects, or repeated meetings over the next review period.

Questions for the Process Champions Committee

- Is this process step clear enough for a new worker to follow without private coaching from a veteran?
- Does the current artifact help make or prove a decision, or does it merely satisfy tradition?
- What exception would be legitimate, and who has authority to approve it?
- How will we know whether the process is improving the work environment rather than simply adding administrative weight?

A Satirical Farewell to Methodology Theater

Before closing, we must pay respectful tribute to Methodology Theater: the grand corporate performance in which a new initiative is announced, everyone receives a binder, the conference room is renamed a war room, and the same unresolved problems are dressed in fresh terminology. Methodology Theater has provided steady employment to slide designers, acronym inventors, and consultants who can say 'organizational transformation' without blinking.

IPI does not deny that outside expertise can help. It denies that expertise becomes valuable simply because it is expensive, laminated, or delivered by someone carrying a laptop bag with aggressive confidence. Real improvement requires understanding the organization, involving the workers, defining practical processes, measuring outcomes, and sustaining discipline after the kickoff snacks are gone.

The restored legacy edition keeps the satire because satire protects the truth from becoming too polite. It reminds us that process improvement is serious work, but the workplace absurdities around it are often hilarious enough to require documentation. If workers can laugh at the dysfunction, they can also begin to name it. If they can name it, they can improve it.

Practical Application

- Define the specific behavior in this area that the organization expects from each role before asking workers to demonstrate compliance.
- Identify the artifact or evidence that proves the behavior occurred; if no evidence exists, decide whether the process needs evidence or whether the requirement is unnecessary ceremony.
- Assign an owner who can answer questions, update guidance, and bring unresolved issues to the Process Champions Committee.
- Measure whether this discipline reduces rework, confusion, delays, defects, or repeated meetings over the next review period.

Questions for the Process Champions Committee

- Is this process step clear enough for a new worker to follow without private coaching from a veteran?
- Does the current artifact help make or prove a decision, or does it merely satisfy tradition?
- What exception would be legitimate, and who has authority to approve it?
- How will we know whether the process is improving the work environment rather than simply adding administrative weight?

Closing Charge to the Reader

The reader who reaches this point should not leave with the desire to copy every template, schedule every committee, or rename every folder by Monday morning. The reader should leave with a more important conviction: process improvement must be intelligent, incremental, measurable, harmonious, empowering, grounded in real work, and aimed at reducing frustration. That was the original promise of IPI, and it remains the promise of this restored legacy edition.

Begin where the pain is visible. Find one broken handoff. Fix one confusing template. Measure one recurring defect pattern. Maintain one improvement long enough for people to trust it. Then repeat. Do not try to boil the ocean, certify the ocean, audit the ocean, and then hire consultants to explain why the ocean resisted change. Start small. Improve honestly. Protect the workers. Respect the process. Document what matters. Learn from what breaks.

The legacy of structured software development is not nostalgia for bureaucracy. It is respect for disciplined work. It is the belief that quality does not happen by accident, that accountability should not be selective, and that real workers deserve processes that make their jobs clearer rather than more frustrating. That is the harmony worth restoring.

Practical Application

- Define the specific behavior in this area that the organization expects from each role before asking workers to demonstrate compliance.
- Identify the artifact or evidence that proves the behavior occurred; if no evidence exists, decide whether the process needs evidence or whether the requirement is unnecessary ceremony.
- Assign an owner who can answer questions, update guidance, and bring unresolved issues to the Process Champions Committee.
- Measure whether this discipline reduces rework, confusion, delays, defects, or repeated meetings over the next review period.

Questions for the Process Champions Committee

- Is this process step clear enough for a new worker to follow without private coaching from a veteran?
- Does the current artifact help make or prove a decision, or does it merely satisfy tradition?
- What exception would be legitimate, and who has authority to approve it?
- How will we know whether the process is improving the work environment rather than simply adding administrative weight?

Implementation Summary

Area	Legacy Discipline	Evidence
Governance	Decision owner identified	PCC minutes or approval log
Artifacts	Required templates controlled	Process Library register
Training	Role impact explained	Training record or quick guide
Audit	Compliance and usefulness checked	Audit checklist and findings
Metrics	Outcome measured over time	Monthly metric trend
Improvement	Corrective action maintained	Closure record and lessons learned

Final Legacy Principles

Legacy Field Observation

Final-word trouble appeared when real workers were thanked for heroics instead of being given a process that made heroics less necessary.

The practical lesson is simple: when ownership combines with accountability, the organization begins paying interest on process debt. That interest appears as rework, political defense, duplicate meetings, unexplained status colors, and heroic weekends that are celebrated publicly and resented privately.

IPI ends where it began: with the real worker, the actual work, and the discipline required to make the workplace more humane and predictable.

Practical Guidance

- Define the normal path for ownership in one page before debating the exception path for three weeks.
- Assign a named owner for accountability, because anonymous ownership is where good processes go to become folklore.
- Create a visible artifact for harmony so teams can inspect the same evidence instead of arguing from memory.
- Measure sustainability only if the result will trigger a decision, correction, or meaningful conversation.
- Review practical pride at a fixed cadence so improvement does not depend on one exhausted champion with a color-coded binder.

Legacy Insert: What Good Looks Like

A good final lesson gives real workers more than gratitude; it gives them a language for protecting their work.

Control	Focus	Minimum Evidence	Review Cadence
Control Point 1	Ownership	Owner named, artifact visible, timing defined, exception path documented	Weekly during pilot; monthly after stabilization
Control Point 2	Accountability	Owner named, artifact visible, timing defined, exception path documented	Weekly during pilot; monthly after stabilization
Control Point 3	Harmony	Owner named, artifact visible, timing defined, exception path documented	Weekly during pilot; monthly after stabilization
Control Point 4	Sustainability	Owner named, artifact visible, timing defined, exception path documented	Weekly during pilot; monthly after stabilization
Control Point 5	Practical Pride	Owner named, artifact visible, timing defined, exception path documented	Weekly during pilot; monthly after stabilization

Mini-Case Scenario

A final improvement effort begins with confidence because the real workers already know where the process is broken. Two months later, the team discovers that ownership was assumed, accountability was informal, and harmony was discussed in a hallway by people who are now on vacation. The project manager asks for accountability. The workers ask for documentation. The sponsor asks why the status was green last week. Everyone looks at the tester, because testers are apparently the archaeologists of bad decisions.

The IPI response is to turn real-worker survival knowledge into a process asset the organization can keep.

Chapter Checklist

#	Checklist Item	Status	Notes
1	The role accountable for the activity is named.	Yes / No / Partial	Evidence or owner
2	The required artifact is defined in plain language.	Yes / No / Partial	Evidence or owner
3	The approval point is visible before downstream work begins.	Yes / No / Partial	Evidence or owner
4	The exception path is documented and time-limited.	Yes / No / Partial	Evidence or owner
5	The measurement has an action threshold.	Yes / No / Partial	Evidence or owner
6	The process can be taught to a new employee without folklore.	Yes / No / Partial	Evidence or owner

IPI Reality Check

A process book without forms is like a methodology consultant without a slide deck: technically possible, but nobody in the conference room believes it exists.

APPENDIX A

Extended Feng Shui Principles for Legacy Process Improvement

The original IPI premise connected software process improvement to workplace harmony. In this restored legacy edition, Feng Shui is not treated as decoration, superstition, or an office-cubicle craft project involving mirrors, bamboo flutes, and a suspicious number of red ribbons. It is used as a metaphor for balance, environment, intention, energy, and the relationship between people and process.

In the legacy process world, the work environment has its own Qi. A project with clear requirements, stable ownership, thoughtful change control, and honest metrics has different energy than a project where everyone is pretending the schedule is realistic because the executive sponsor smiled during the kickoff meeting.

IPI Reality Check

Negative Qi is not always mystical. Sometimes it is a requirements document named Final_Final_ReallyFinal_v7.doc saved on someone's desktop.

Unity / Teamwork

The energy of unity shows up when project participants understand that software delivery is not a relay race where each function throws work over a wall and calls it collaboration.

Legacy application: make executive support visible through clear sponsorship, decision authority, escalation paths, and evidence that leadership protects the process rather than merely announces it.

Balance / Customer Neutrality

Balance means the customer is heard without allowing every midstream idea to become an emergency requirement disguised as a clarification.

Legacy application: make the customer role explicit by documenting input points, approval responsibilities, feedback windows, and the evidence that business needs were heard before downstream work began.

Prosperity / Real Worker Contribution

Prosperity comes when the organization stops treating the people closest to the work as the last people to consult.

Legacy application: rotate worker representation with purpose: define the term, the responsibilities, the feedback path, and the way frontline knowledge becomes process improvement.

Reputation / Process Training

A company's reputation is protected when people know how work should be performed before they are blamed for performing it incorrectly.

Legacy application: connect training to performance by defining what must be learned, who must learn it, how competency is verified, and how updates are communicated when the process changes.

Partnership / Process Library

Partnership improves when everyone can find the same approved artifact instead of relying on folklore, inbox archaeology, and the one employee who knows where the template lives.

Legacy application: treat the process library as controlled knowledge by assigning ownership, version rules, review cycles, retirement rules, and a location workers can trust.

Creativity / Process Project Management

Creativity is not the enemy of structure. Creativity needs boundaries so it can produce value instead of improvisational chaos.

Legacy application: give the process project manager a visible charter, work plan, dependency map, issue path, and authority to coordinate improvement without becoming another bottleneck.

Progress / Business Analysis

Progress begins when current-state confusion is made visible and future-state discipline is made practical.

Legacy application: use analysis to map pain points into practical process changes, then confirm that the improvement reduces confusion, rework, or delay.

Promotion / Auditing

Promotion is the recognition of disciplined work. Process audits should reveal learning, not merely announce guilt.

Legacy application: use audits to reveal whether the described process matches the performed process, then convert findings into corrective action rather than fear.

Longevity / Configuration Management

Longevity is created when the organization knows what changed, who approved it, where it lives, and how to restore order when the release turns dramatic.

Legacy application: make configuration control protect the integrity of releases, baselines, documents, and environments without turning every small change into a festival of approvals.

Health / Metrics

Healthy measurement separates useful signals from decorative numbers that look important in a status meeting but cannot improve anything.

Legacy application: define metrics that expose process health, support decisions, and identify imbalance before the organization mistakes motion for improvement.

Holistic / IPI Advising

The wholistic view connects intelligence, surroundings, and values so that process improvement serves people rather than exhausting them.

Legacy application: use the IPI advisor role to connect people, process, surroundings, and values so that discipline remains humane, practical, and sustainable.

Feng Shui-to-Process Reflections:

Use these reflections before introducing a new process, revising an old one, or attempting to rescue a project that has already acquired the unmistakable scent of schedule denial. Each reflection connects Feng Shui balance to practical IPI process discipline so the team can locate blocked energy, misplaced accountability, and process objects that need to be restored to their proper place.

Reflection 1 — Workplace Flow

Use this reflection when work is moving, but not flowing. In Feng Shui terms, the path is blocked; in IPI terms, handoffs, decisions, and artifacts are not moving cleanly through the SDLC.

- Where does work wait, loop backward, or disappear before the next role can act?
- Which handoff creates the most rework, delay, or private escalation?
- What artifact should make the handoff visible but is missing, outdated, or ignored?
- Which role must own the transition so the process does not depend on hallway memory?
- What evidence will show that flow has improved instead of merely being announced?

Reflection 2 — Role Balance

Use this reflection when one person, role, or department is absorbing frustration that properly belongs to the process.

- Which role is carrying hidden work that should be shared, documented, or governed?
- Where are responsibilities overlapping, duplicated, or quietly abandoned?
- Which decision rights are unclear enough to create delay or conflict?
- What simple role definition would reduce finger pointing without creating bureaucracy?
- How will the PCC know that responsibility is now balanced across the workgroup?

Reflection 3 — Artifact Harmony

Use this reflection when documents, templates, records, and evidence are scattered like process furniture after a failed office move.

- Which artifact is missing, duplicated, obsolete, or stored in the wrong location?
- Which document do real workers trust, and which one exists only for audit theater?
- What artifact should be retired so workers stop choosing between competing truths?
- Who owns the artifact, version, review date, and update path?
- What placement rule will keep the right artifact easy to find and hard to misuse?

Reflection 4 — Accountability Without Fear

Use this reflection when accountability has become blame, silence, or defensive status reporting.

- What decision, approval, or result currently lacks a clear owner?

- Where are people avoiding transparency because accountability feels punitive?
- What evidence should prove ownership without turning the process into surveillance?
- How can the process make responsibility visible before failure occurs?
- What change would create trust while still preserving discipline?

Reflection 5 — Process Energy Leaks

Use this reflection when the team is busy, tired, and still unable to explain why the same problems keep returning.

- Which meeting, report, template, or approval consumes energy without improving work?
- Where does rework repeatedly appear after the team believed the issue was settled?
- Which process step creates more confusion than clarity?
- What lightweight control would stop the leak without adding unnecessary weight?
- How will the team measure recovered time, reduced rework, or improved clarity?

Reflection 6 — Requirements Stability

Use this reflection when requirements behave like moving furniture in a dark room: everyone keeps bumping into something that was not supposed to move.

- Which requirement source is authoritative, and how does the team know?
- What requirement change has entered the project without proper impact review?
- Where are approvals unclear, late, or based on conversation instead of evidence?
- What baseline or traceability control would restore order?

- What proof will show that requirements are stable enough for downstream work?

Reflection 7 — Quality Balance

Use this reflection when quality is treated as the tester's burden rather than the organization's shared responsibility.

- Where should quality be built earlier into requirements, design, or review activity?
- Which defect pattern reveals a process weakness upstream?
- What review, checklist, or verification point would prevent repeated escape defects?
- Who owns quality evidence before the product reaches final test?
- What quality signal will show that prevention is improving, not just detection?

Reflection 8 — Measurement With Meaning

Use this reflection when metrics look impressive but no one can name a decision they helped make.

- Which metric is currently decorative rather than actionable?
- What question should the measurement answer for workers, managers, or customers?
- What threshold should trigger review, correction, or escalation?
- Who validates the data so the metric does not become fiction with decimal points?
- What decision will improve because this measurement exists?

Reflection 9 — Training and Knowledge Flow

Use this reflection when the process exists, but workers learn it through folklore, rescue calls, and the one veteran who knows where everything is hidden.

- Which process step requires private coaching because formal guidance is unclear?

- What training is missing, outdated, or delivered too late to help the work?
- Which role needs job aids, examples, or refreshers to perform consistently?
- How will new workers learn the process without inheriting bad habits?
- What evidence shows that knowledge has moved from individuals into the organization?

Reflection 10 — Process Library Order

Use this reflection when the process library has become a shared-drive attic filled with old templates, mystery folders, and documents named final-final-this-time.

- Can a worker find the correct current process asset in less than two minutes?
- Which assets should be archived, renamed, consolidated, or retired?
- What ownership and review cycle keeps the library current?
- Which document naming, versioning, or folder rule would reduce confusion immediately?
- What usage evidence will show that the process library is trusted and alive?

Reflection 11 — Change Control Balance

Use this reflection when change is either uncontrolled chaos or a ceremonial traffic jam.

- Which changes require formal control, and which can be handled through lighter guidance?
- Where does the current change process delay necessary work without reducing risk?
- What impact areas must be reviewed: requirements, design, testing, schedule, cost, training, support, or documentation?

- Who has authority to approve, reject, defer, or escalate the change?
- What evidence will show that change is controlled without freezing progress?

Reflection 12 — Sustained Harmony

Use this reflection when the improvement has been announced, but maintenance, ownership, training, and measurement have not yet been made real.

- What must be maintained so the improvement does not fade after the kickoff enthusiasm ends?
- Which lesson learned should become a process update, checklist item, or training point?
- What audit or review cadence will confirm the new process is being followed?
- Who owns continued improvement after the initial champion becomes busy with real work again?
- What sign will show that the workplace has achieved lasting harmony rather than temporary compliance?

IPI Reality Check

Do not confuse harmony with silence. In many organizations, the room is quiet because everyone has already updated their resume.

APPENDIX B

IPI Self-Assessment Questionnaire

This questionnaire is designed to help an organization identify whether it is operating with disciplined process energy or surviving through personality, heroics, memory, and recurring miracles.

Scoring Key

Use 1 for Not True, 2 for Rarely True, 3 for Sometimes True, 4 for Mostly True, and 5 for Consistently True.

- Executive Sponsorship: roles and responsibilities are clearly understood before execution begins. Score: 1 2 3 4 5
- Executive Sponsorship: required artifacts are defined, accessible, and version controlled. Score: 1 2 3 4 5
- Executive Sponsorship: deviations are documented rather than hidden in email threads. Score: 1 2 3 4 5
- Executive Sponsorship: the people doing the real work have input into process improvement. Score: 1 2 3 4 5
- Executive Sponsorship: metrics are reviewed for decision-making, not just presentation decoration. Score: 1 2 3 4 5
- Customer Involvement: roles and responsibilities are clearly understood before execution begins. Score: 1 2 3 4 5
- Customer Involvement: required artifacts are defined, accessible, and version controlled. Score: 1 2 3 4 5
- Customer Involvement: deviations are documented rather than hidden in email threads. Score: 1 2 3 4 5
- Customer Involvement: the people doing the real work have input into process improvement. Score: 1 2 3 4 5
- Customer Involvement: metrics are reviewed for decision-making, not just presentation decoration. Score: 1 2 3 4 5

- Requirements Discipline: roles and responsibilities are clearly understood before execution begins. Score: 1 2 3 4 5
- Requirements Discipline: required artifacts are defined, accessible, and version controlled. Score: 1 2 3 4 5

IPI Reality Check

If every answer is a 5, either congratulations are in order or the questionnaire was completed by the same committee that writes the annual executive status report.

- Requirements Discipline: deviations are documented rather than hidden in email threads. Score: 1 2 3 4 5
- Requirements Discipline: the people doing the real work have input into process improvement. Score: 1 2 3 4 5
- Requirements Discipline: metrics are reviewed for decision-making, not just presentation decoration. Score: 1 2 3 4 5
- Project Planning: roles and responsibilities are clearly understood before execution begins. Score: 1 2 3 4 5
- Project Planning: required artifacts are defined, accessible, and version controlled. Score: 1 2 3 4 5
- Project Planning: deviations are documented rather than hidden in email threads. Score: 1 2 3 4 5
- Project Planning: the people doing the real work have input into process improvement. Score: 1 2 3 4 5
- Project Planning: metrics are reviewed for decision-making, not just presentation decoration. Score: 1 2 3 4 5
- Configuration Management: roles and responsibilities are clearly understood before execution begins. Score: 1 2 3 4 5
- Configuration Management: required artifacts are defined, accessible, and version controlled. Score: 1 2 3 4 5
- Configuration Management: deviations are documented rather than hidden in email threads. Score: 1 2 3 4 5

- Configuration Management: the people doing the real work have input into process improvement. Score: 1 2 3 4 5
- Configuration Management: metrics are reviewed for decision-making, not just presentation decoration. Score: 1 2 3 4 5
- Testing and Quality: roles and responsibilities are clearly understood before execution begins. Score: 1 2 3 4 5
- Testing and Quality: required artifacts are defined, accessible, and version controlled. Score: 1 2 3 4 5
- Testing and Quality: deviations are documented rather than hidden in email threads. Score: 1 2 3 4 5
- Testing and Quality: the people doing the real work have input into process improvement. Score: 1 2 3 4 5
- Testing and Quality: metrics are reviewed for decision-making, not just presentation decoration. Score: 1 2 3 4 5
- Release Readiness: roles and responsibilities are clearly understood before execution begins. Score: 1 2 3 4 5
- Release Readiness: required artifacts are defined, accessible, and version controlled. Score: 1 2 3 4 5
- Release Readiness: deviations are documented rather than hidden in email threads. Score: 1 2 3 4 5
- Release Readiness: the people doing the real work have input into process improvement. Score: 1 2 3 4 5
- Release Readiness: metrics are reviewed for decision-making, not just presentation decoration. Score: 1 2 3 4 5
- Process Library: roles and responsibilities are clearly understood before execution begins. Score: 1 2 3 4 5
- Process Library: required artifacts are defined, accessible, and version controlled. Score: 1 2 3 4 5
- Process Library: deviations are documented rather than hidden in email threads. Score: 1 2 3 4 5
- Process Library: the people doing the real work have input into process improvement. Score: 1 2 3 4 5

- Process Library: metrics are reviewed for decision-making, not just presentation decoration. Score: 1 2 3 4 5
- Training: roles and responsibilities are clearly understood before execution begins. Score: 1 2 3 4 5
- Training: required artifacts are defined, accessible, and version controlled. Score: 1 2 3 4 5
- Training: deviations are documented rather than hidden in email threads. Score: 1 2 3 4 5
- Training: the people doing the real work have input into process improvement. Score: 1 2 3 4 5
- Training: metrics are reviewed for decision-making, not just presentation decoration. Score: 1 2 3 4 5
- Metrics: roles and responsibilities are clearly understood before execution begins. Score: 1 2 3 4 5
- Metrics: required artifacts are defined, accessible, and version controlled. Score: 1 2 3 4 5
- Metrics: deviations are documented rather than hidden in email threads. Score: 1 2 3 4 5
- Metrics: the people doing the real work have input into process improvement. Score: 1 2 3 4 5
- Metrics: metrics are reviewed for decision-making, not just presentation decoration. Score: 1 2 3 4 5
- Audits: roles and responsibilities are clearly understood before execution begins. Score: 1 2 3 4 5
- Audits: required artifacts are defined, accessible, and version controlled. Score: 1 2 3 4 5
- Audits: deviations are documented rather than hidden in email threads. Score: 1 2 3 4 5
- Audits: the people doing the real work have input into process improvement. Score: 1 2 3 4 5
- Audits: metrics are reviewed for decision-making, not just presentation decoration. Score: 1 2 3 4 5

- Lessons Learned: roles and responsibilities are clearly understood before execution begins. Score: 1 2 3 4 5
- Lessons Learned: required artifacts are defined, accessible, and version controlled. Score: 1 2 3 4 5
- Lessons Learned: deviations are documented rather than hidden in email threads. Score: 1 2 3 4 5
- Lessons Learned: the people doing the real work have input into process improvement. Score: 1 2 3 4 5
- Lessons Learned: metrics are reviewed for decision-making, not just presentation decoration. Score: 1 2 3 4 5

APPENDIX C

Process Champions Committee Charter and Operating Forms

The following forms are intended as legacy working tools: printable, auditable, and simple enough that a real worker can use them without needing a certification, a consultant, or a laminated methodology decoder ring.

PCC Charter Form

IPI Reality Check

A form should clarify work. If completing the form requires a meeting to explain the form, the form has become the work.

Purpose	________________________________
Executive Sponsor	________________________________
Standing Members	________________________________
Rotating Representatives	________________________________
Decision Rights	________________________________
Meeting Cadence	________________________________
Escalation Path	________________________________
Repository Location	________________________________
Approval Date	________________________________

Rotating Representative Nomination Form

Workgroup	______________________________
Nominee	______________________________
Term Start	______________________________
Term End	______________________________
Required Training Completed	______________________________
Supervisor Approval	______________________________
Representative Responsibilities	______________________________
Known Constraints	______________________________

PCC Meeting Agenda

Meeting Date	______________________________________
Facilitator	______________________________________
Attendees	______________________________________
Open Process Requests	______________________________________
Pilots in Progress	______________________________________
Audit Findings	______________________________________
Metrics Reviewed	______________________________________
Decisions Made	______________________________________
Action Items	______________________________________

PCC Decision Log

Decision ID	______________________________
Decision Summary	______________________________
Alternatives Considered	______________________________
Impact	______________________________
Owner	______________________________
Due Date	______________________________
Approval	______________________________
Communication Plan	______________________________

APPENDIX D

Four-Tier Process Hierarchy Forms

The following forms are intended as legacy working tools: printable, auditable, and simple enough that a real worker can use them without needing a certification, a consultant, or a laminated methodology decoder ring.

Tier 1 Policy Definition Form

Policy Name	______________________________
Business Reason	______________________________
Policy Owner	______________________________
Required Compliance	______________________________
Affected Workgroups	______________________________
Effective Date	______________________________
Review Cycle	______________________________

IPI Reality Check

A form should clarify work. If completing the form requires a meeting to explain the form, the form has become the work.

Tier 2 Process Definition Form

Process Name	________________________
Trigger	________________________
Inputs	________________________
Steps	________________________
Outputs	________________________
Roles	________________________
Exit Criteria	________________________
Exceptions	________________________

Tier 3 Procedure Form

Procedure Name	________________________________
Tool Used	________________________________
Step-by-Step Instructions	________________________________
Required Screenshots	________________________________
Validation Method	________________________________
Common Errors	________________________________

Tier 4 Template Control Form

Template Name	________________________________
Artifact Type	________________________________
Required Fields	________________________________
Optional Fields	________________________________
Storage Location	________________________________
Version	________________________________
Approver	________________________________

APPENDIX E

Find, Fix, Measure, Maintain Workbook

The following forms are intended as legacy working tools: printable, auditable, and simple enough that a real worker can use them without needing a certification, a consultant, or a laminated methodology decoder ring.

Find Phase Gap Card

Gap ID	______________________________
Observed Problem	______________________________
Negative Qi / Frustration	______________________________
Affected Role	______________________________
Evidence	______________________________
Priority	______________________________
Suggested Owner	______________________________

IPI Reality Check

A form should clarify work. If completing the form requires a meeting to explain the form, the form has become the work.

Fix Phase Pilot Plan

Pilot Name	______________________________
Process Change	______________________________
Pilot Project	______________________________
Participants	______________________________
Start Date	______________________________
End Date	______________________________
Risks	______________________________
Pilot Success Criteria	______________________________

Measure Phase Results Form

Pilot Name	____________________
Process Change	____________________
Pilot Project	____________________
Participants	____________________
Start Date	____________________
End Date	____________________
Risks	____________________
Pilot Success Criteria	____________________

Maintain Phase Institutionalization Checklist

Training Completed	________________________________
Repository Updated	________________________________
Audit Criteria Updated	________________________________
Metrics Added	________________________________
Owner Assigned	________________________________
Next Review Date	________________________________

APPENDIX F

Legacy SDLC Forms and Checklists

The following forms are intended as legacy working tools: printable, auditable, and simple enough that a real worker can use them without needing a certification, a consultant, or a laminated methodology decoder ring.

Project Definition Checklist

Business Need	___________________________
Sponsor	___________________________
Scope Statement	___________________________
Known Constraints	___________________________
Initial Risks	___________________________
Funding Approval	___________________________
Go/ No-Go Decision	___________________________

IPI Reality Check

A form should clarify work. If completing the form requires a meeting to explain the form, the form has become the work.

Requirements Sign-Off Form

Requirement Set	______________________________
Business Owner	______________________________
Analyst	______________________________
Review Date	______________________________
Open Questions	______________________________
Approved / Rejected	______________________________
Change Control Reminder	______________________________

Design Review Checklist

Architecture Reviewed	____
Interface Impact	____
Data Impact	____
Security Considerations	____
Operational Impact	____
Approval	____

Test Readiness Review

Test Plan	____________________
Test Data	____________________
Environment	____________________
Entry Criteria	____________________
Defect Process	____________________
Traceability	____________________
Readiness Decision	____________________

Deployment Readiness Checklist

Release Notes	____
Backout Plan	____
Approvals	____
Configuration Items	____
Known Issues	____
Support Handoff	____
Deployment Window	____

APPENDIX G

Process Audit Forms

The following forms are intended as legacy working tools: printable, auditable, and simple enough that a real worker can use them without needing a certification, a consultant, or a laminated methodology decoder ring.

Process Audit Plan

Audit Scope	______________________________
Audit Period	______________________________
Auditor	______________________________
Process Areas	______________________________
Artifacts Requested	______________________________
Interviewees	______________________________
Audit Schedule	______________________________

Artifact Evidence Checklist

Artifact	____
Required Version	____
Repository Location	____
Owner	____
Present	____
Complete	____
Notes	____

Audit Finding Form

Finding ID	________________________________
Process Area	________________________________
Observation	________________________________
Evidence	________________________________
Severity	________________________________
Root Cause	________________________________
Corrective Action	________________________________
Due Date	________________________________

Corrective Action Verification

Finding ID	____
Action Taken	____
Evidence Reviewed	____
Verification Date	____
Status	____
Auditor Sign-Off	____

APPENDIX H

Training and Process Memory Forms

The following forms are intended as legacy working tools: printable, auditable, and simple enough that a real worker can use them without needing a certification, a consultant, or a laminated methodology decoder ring.

Training Needs Analysis

Role	________________________________
Required Process Knowledge	________________________________
Current Gap	________________________________
Training Method	________________________________
Due Date	________________________________
Owner	________________________________

Training Attendance Record

Course	________________________________
Date	________________________________
Instructor	________________________________
Participant	________________________________
Workgroup	________________________________
Completion Status	________________________________
Score	________________________________

Process Memory Capture Form

Topic	________________________________
Experienced Worker	________________________________
Known Shortcuts	________________________________
Known Risks	________________________________
Historical Lessons	________________________________
Recommended Update	________________________________

New Employee Process Orientation Checklist

Repository Tour	____________________________
SDLC Overview	____________________________
Template Training	____________________________
Change Control	____________________________
Audit Expectations	____________________________
Metrics Overview	____________________________

APPENDIX I

Process Library and Artifact Control Register

The following forms are intended as legacy working tools: printable, auditable, and simple enough that a real worker can use them without needing a certification, a consultant, or a laminated methodology decoder ring.

Process Asset Register

Asset ID	________________________________
Asset Name	________________________________
Type	________________________________
Owner	________________________________
Version	________________________________
Effective Date	________________________________
Repository Path	________________________________
Review Date	________________________________

Artifact Retirement Form

Artifact Name	____
Reason for Retirement	____
Replacement Artifact	____
Impact	____
Approval	____
Archive Location	____

Repository Health Checklist

Duplicate Files Removed	____
Naming Standards Applied	____
Broken Links Fixed	____
Old Versions Archived	____
Access Reviewed	____

Template Change Request

Template	______________________________
Requested Change	______________________________
Reason	______________________________
Affected Roles	______________________________
Approver	______________________________
Implementation Date	______________________________

APPENDIX J

Metrics Dictionary and Measurement Forms

The following forms are intended as legacy working tools: printable, auditable, and simple enough that a real worker can use them without needing a certification, a consultant, or a laminated methodology decoder ring.

Metrics Dictionary Entry

Metric Name	______________________________
Purpose	______________________________
Definition	______________________________
Formula	______________________________
Data Source	______________________________
Collection Frequency	______________________________
Owner	______________________________
Interpretation	______________________________

Measurement Collection Sheet

Metric	______________________________
Period	______________________________
Raw Data	______________________________
Calculated Value	______________________________
Trend	______________________________
Comments	______________________________
Action Required	______________________________

Decorative Number Detection Form

Reported Number	________________________________
Decision It Supports	________________________________
Data Source	________________________________
Confidence Level	________________________________
Potential Misuse	________________________________
Keep / Retire	________________________________

Process Health Dashboard Notes

Area	______________________________
Green Criteria	______________________________
Yellow Criteria	______________________________
Red Criteria	______________________________
Current Status	______________________________
Action	______________________________

APPENDIX K

Change Control, Waiver, and Lessons Learned Forms

The following forms are intended as legacy working tools: printable, auditable, and simple enough that a real worker can use them without needing a certification, a consultant, or a laminated methodology decoder ring.

Change Request Form

Change ID	______________________
Requester	______________________
Description	______________________
Business Reason	______________________
Impact	______________________
Risk	______________________
Decision	______________________
Approval	______________________

Process Waiver Request

Waiver ID	______________________________
Process	______________________________
Reason for Waiver	______________________________
Duration	______________________________
Risk Accepted	______________________________
Approver	______________________________
Expiration Date	______________________________

Lessons Learned Form

Project	______________________________
What Worked	______________________________
What Failed	______________________________
Root Cause	______________________________
Recommended Process Update	______________________________
Owner	______________________________
Due Date	______________________________

Post-Implementation Review

Deployment Result	______________________________
Defects Introduced	______________________________
Customer Impact	______________________________
Support Issues	______________________________
Process Improvements	______________________________
Final Recommendation	______________________________

APPENDIX L

Glossary of Legacy IPI Terms

Accountability

The disciplined agreement that a role, artifact, decision, or outcome belongs to someone specific rather than to the foggy kingdom known as "the team."

Artifact

A document, record, template, log, diagram, or other evidence that work was performed and can be reviewed without relying on memory.

Baseline

The approved point of reference before everyone begins remembering the plan differently.

Change Control

The practice of deciding whether a change should happen before pretending it already did.

Configuration Management

The discipline of knowing what version exists, where it lives, who changed it, and why the production environment is not the place to discover confusion.

CMMI

A process improvement model respected for discipline and feared for its ability to make average workers reach for aspirin when implemented without translation.

Current State

The way work actually happens, usually discovered after comparing official procedures to the behavior of people trying to survive the deadline.

Decorative Metric

A number used because it looks executive-friendly rather than because it supports a decision.

Find

The IPI phase where gaps, friction, negative Qi, and process pain are identified honestly.

Fix

The IPI phase where a realistic improvement is designed and piloted before being declared universal law.

Measure

The IPI phase where evidence is collected to determine whether the fix worked or merely produced prettier documentation.

Maintain

The IPI phase where the organization keeps the improvement alive after the kickoff enthusiasm fades.

Negative Qi

The workplace energy produced by unclear ownership, missing artifacts, unmanaged change, political silence, and vague requirements.

PCC

Process Champions Committee; the body that translates process improvement into practical workgroup discipline.

Pilot

A controlled trial used to prove whether a process works before unleashing it on everyone like a memo from the policy department.

Process Library

The controlled repository where approved process assets live, ideally not inside someone's inbox.

Real Workers

The people who execute the work, absorb the contradictions, and know which official process steps are theater.

Repository

The place where process assets should be stored; not to be confused with a shared drive landfill.

Waiver

Documented permission to deviate from a process for a defined reason, duration, and risk.

Audit Trail

Audit Trail is a legacy process term used to preserve clarity, ownership, evidence, and repeatability across the software development life cycle. In IPI practice, the term should be defined simply enough that the people doing the work can apply it without stopping the project to translate methodology language.

Business Rule

Business Rule is a legacy process term used to preserve clarity, ownership, evidence, and repeatability across the software development life cycle. In IPI practice, the term should be defined simply enough that the people doing the work can apply it without stopping the project to translate methodology language.

Customer Sign-Off

Customer Sign-Off is a legacy process term used to preserve clarity, ownership, evidence, and repeatability across the software development life cycle. In IPI practice, the term should be defined simply enough that the people doing the work can apply it without stopping the project to translate methodology language.

Defect Leakage

Defect Leakage is a legacy process term used to preserve clarity, ownership, evidence, and repeatability across the software development life cycle. In IPI practice, the term should be defined simply enough that the people doing the work can apply it without stopping the project to translate methodology language.

Entry Criteria

Entry Criteria is a legacy process term used to preserve clarity, ownership, evidence, and repeatability across the software development life cycle. In IPI practice, the term should be defined simply enough that the people doing the work can apply it without stopping the project to translate methodology language.

Exit Criteria

Exit Criteria is a legacy process term used to preserve clarity, ownership, evidence, and repeatability across the software development life cycle. In IPI practice, the term should be defined simply enough that the people doing the work can apply it without stopping the project to translate methodology language.

Gap Analysis

Gap Analysis is a legacy process term used to preserve clarity, ownership, evidence, and repeatability across the software development life cycle. In IPI practice, the term should be defined simply enough that the people doing the work can apply it without stopping the project to translate methodology language.

Impact Analysis

Impact Analysis is a legacy process term used to preserve clarity, ownership, evidence, and repeatability across the software development life cycle. In IPI practice, the term should be defined simply enough that the people doing the work can apply it without stopping the project to translate methodology language.

Lessons Learned

Lessons Learned is a legacy process term used to preserve clarity, ownership, evidence, and repeatability across the software development life cycle. In IPI practice, the term should be defined simply enough that the people doing the work can apply it without stopping the project to translate methodology language.

Process Owner

Process Owner is a legacy process term used to preserve clarity, ownership, evidence, and repeatability across the software development life cycle. In IPI practice, the term should be defined simply enough that the people doing the work can apply it without stopping the project to translate methodology language.

Quality Gate

Quality Gate is a legacy process term used to preserve clarity, ownership, evidence, and repeatability across the software development life cycle. In IPI practice, the term should be defined simply enough that the people doing the work can apply it without stopping the project to translate methodology language.

Release Package

Release Package is a legacy process term used to preserve clarity, ownership, evidence, and repeatability across the software development life cycle. In IPI practice, the term should be defined simply enough that the people doing the work can apply it without stopping the project to translate methodology language.

Requirements Traceability

Requirements Traceability is a legacy process term used to preserve clarity, ownership, evidence, and repeatability across the software development life cycle. In IPI practice, the term should be defined simply enough that the people doing the work can apply it without stopping the project to translate methodology language.

Risk Register

Risk Register is a legacy process term used to preserve clarity, ownership, evidence, and repeatability across the software development life cycle. In IPI practice, the term should be defined simply enough that the people doing the work can apply it without stopping the project to translate methodology language.

Root Cause

Root Cause is a legacy process term used to preserve clarity, ownership, evidence, and repeatability across the software development life cycle. In IPI practice, the term should be defined simply enough that the people doing the work can apply it without stopping the project to translate methodology language.

Standard Work

Standard Work is a legacy process term used to preserve clarity, ownership, evidence, and repeatability across the software development life cycle. In IPI practice, the term should be defined simply enough that the people doing the work can apply it without stopping the project to translate methodology language.

Template

Template is a legacy process term used to preserve clarity, ownership, evidence, and repeatability across the software development life cycle. In IPI practice, the term should be defined simply enough that the people doing the work can apply it without stopping the project to translate methodology language.

Training Matrix

Training Matrix is a legacy process term used to preserve clarity, ownership, evidence, and repeatability across the software development life cycle. In IPI practice, the term should be defined simply enough that the people doing the work can apply it without stopping the project to translate methodology language.

Version Number

Version Number is a legacy process term used to preserve clarity, ownership, evidence, and repeatability across the software development life cycle. In IPI practice, the term should be defined simply enough that the people doing the work can apply it without stopping the project to translate methodology language.

Workgroup Representative

Workgroup Representative is a legacy process term used to preserve clarity, ownership, evidence, and repeatability across the software development life cycle. In IPI practice, the term should be defined simply enough that the people doing the work can apply it without stopping the project to translate methodology language.

APPENDIX M

Satirical Field Notes from the Process Trenches

Field Note 1: The Ancient Pattern Repeats

Every legacy IT shop has a story that begins with a noble initiative and ends with a spreadsheet nobody opens. The purpose of these field notes is not to mock discipline; it is to mock the strange organizational habit of confusing motion with improvement.

- If the process only exists in PowerPoint, it is not institutionalized; it is illustrated.
- If the only person who understands the process is leaving Friday, the process is not mature; it is endangered.
- If a metric cannot influence a decision, it is office decoration with decimal points.
- If every exception becomes normal, the waiver process has become the process.
- If the template is harder to complete than the work it describes, the template has achieved management status.

IPI Reality Check

The goal is not to eliminate humor from process improvement. The goal is to eliminate the conditions that make gallows humor the only survival tool left.

Field Note 2: The Binder That No One Opens

In the old world, every serious process initiative eventually produced a binder large enough to injure a project coordinator. The binder was usually labeled, tabbed, approved, shelved, and then silently abandoned.

- A binder is not a process unless people actually use it.
- A process stored on a shelf has achieved physical existence, not operational value.
- If the binder is updated only before audits, it is not a guide; it is a costume.
- If real workers created their own unofficial shortcut, the binder has already lost authority.
- The best process document is the one that helps the worker before the crisis, not after the blame meeting.

Field Note 3: The Spreadsheet of Eternal Mystery

Every department has at least one spreadsheet that no one fully understands but everyone is afraid to delete. It contains color codes, hidden tabs, formulas written by a departed analyst, and the quiet fear that it may somehow control production.

- If no one knows who owns the spreadsheet, it is not a tool; it is folklore.
- Hidden columns are where accountability goes to nap.
- If the spreadsheet contradicts the official process, the team will follow whichever one causes less pain.
- A useful tracker clarifies work; a mysterious tracker creates ritual behavior.
- When a spreadsheet becomes mission-critical, it deserves governance, backup, and a real owner.

Field Note 4: The Consultant Carousel

The consultant carousel begins when one firm explains what the last firm did wrong, then produces a new slide deck with the same diagrams in a different color palette.

- A consultant who never interviews real workers is diagnosing the patient from the parking lot.
- A methodology cannot be "implemented" by people who do not understand the culture.
- If the recommendation requires a second consulting contract to explain the first contract, beware.
- Consultants should transfer capability, not create dependency.
- The best outside help makes the inside team stronger after the consultants leave.

Field Note 5: The Meeting Before the Meeting

The meeting before the meeting exists to prepare for the meeting where no decision will be made because the real decision-maker is waiting for a summary from someone who attended neither meeting.

- If a meeting exists only to prepare for another meeting, it should produce a clear decision path.
- A meeting without a decision, owner, or next action is workplace weather.
- The more people invited, the less likely anyone feels accountable.
- Meeting minutes are not progress unless they capture commitments.
- The most valuable meeting is often the one that gets canceled because the process already made the answer clear.

Field Note 6: The Sacred Signoff Sheet

The signoff sheet once symbolized clarity, approval, and responsibility. In dysfunctional hands, it becomes a ceremonial document passed around after everyone has already forgotten what was approved.

- A signature without understanding is not approval; it is handwriting.

- Signoff should confirm agreement, not create surprise.
- If approval occurs after the work is complete, the process is performing archaeology.
- Every signoff should identify what is being approved, by whom, and with what authority.
- A good signoff protects both the project and the people responsible for delivering it.

Field Note 7: The Phantom Process Owner

The phantom process owner appears on documents, org charts, and audit records but cannot be found when the process breaks. This role is often assigned to someone "temporarily," which in corporate time means forever.

- Every process needs an owner with actual authority.
- Ownership without time, resources, or decision rights is symbolic accountability.
- If everyone owns the process, no one maintains it.
- A process owner should know when the process is failing in the real world.
- The owner's job is not to defend the process; it is to improve it.

Field Note 8: The Template That Ate the Project

A good template helps workers think clearly. A bad template turns a two-hour task into a three-day documentation pilgrimage.

- Templates should reduce confusion, not manufacture paperwork.
- If workers spend more time formatting the template than understanding the work, the template has become the work.
- Required fields should be required for a reason.
- A template should match the size and risk of the project.
- The best templates guide judgment without replacing common sense.

Field Note 9: The Metric With No Decision Attached

Some metrics exist only because someone once asked for "visibility." They are collected, formatted, colored, trended, and ignored with impressive consistency.

- Every metric should answer a real management or delivery question.
- If a metric cannot trigger action, it is decorative.
- Metrics should reveal truth, not protect reputations.
- A dashboard should help the team steer, not merely admire the scenery.
- The best measurements improve decisions before problems become emergencies.

Field Note 10: The Waiver That Became Standard Practice

The waiver process begins as a controlled exception and slowly becomes the preferred delivery method for teams that have discovered the process is too heavy to survive contact with reality.

- A waiver should be rare, justified, documented, and temporary.
- If the same waiver is approved repeatedly, the process needs review.
- Exceptions reveal where the official process does not fit the work.
- A waiver without expiration is a process escape tunnel.
- The goal is not to eliminate flexibility; the goal is to prevent flexibility from becoming chaos.

Field Note 11: The Shared Drive Archaeology Expedition

Before modern repositories and search tools, there was the shared drive: a vast digital attic where final documents, final-final documents, and truly-final-this-time documents lived together in uneasy peace.

- A process asset that cannot be found cannot be followed.
- Folder structure is governance in disguise.

- Duplicate documents create duplicate truths.
- If workers rely on memory to find artifacts, the repository has failed.
- A healthy process library should make the correct document easier to find than the wrong one.

Field Note 12: The Requirements Document Nobody Read

The requirements document was reviewed, approved, baselined, and ignored until testing began. Then everyone suddenly became a passionate reader.

- Requirements should be understandable before they are approved.
- A long document is not automatically a complete document.
- If testers find the first real ambiguity, the review process failed.
- Requirements should connect business intent to development and test activity.
- The best requirements prevent arguments later by forcing clarity early.

Field Note 13: The Traceability Matrix of Doom

The traceability matrix is one of the most powerful tools in structured software development, provided it is not treated like a punishment invented by auditors.

- Traceability should prove alignment, not create busywork.
- Each requirement should connect to design, development, test, and acceptance evidence.
- A missing trace is an early warning sign, not merely an audit defect.
- Traceability is easiest when built throughout the lifecycle, not reconstructed at the end.

- A good matrix tells the story of the project's logic from need to solution.

Field Note 14: The Configuration Management Treasure Hunt

Configuration management is where many teams discover that "the latest version" is a phrase of hope, not evidence.

- Version control must apply to documents, code, test assets, and release materials.
- A baseline should be known, protected, and recoverable.
- If two people are working from different versions, the process has already split.
- Configuration management prevents accidental creativity during release.
- Control is not bureaucracy when it protects the integrity of the product.

Field Note 15: The Process Library Labyrinth

The process library was created to help workers, but over time it became a maze of folders, outdated PDFs, retired templates, and documents named after committee decisions nobody remembers.

- A process library should be curated, not merely accumulated.
- Retired assets should be clearly separated from current assets.
- Every document should have an owner, version, purpose, and review date.
- A library that contains everything without structure teaches workers to trust nothing.
- The process librarian is not a clerk; the librarian protects organizational memory.

Field Note 16: The Audit Panic Ritual

Audit panic begins when leadership suddenly remembers that process compliance exists. Teams scramble, documents appear, and everyone becomes deeply interested in naming conventions.

- Audit readiness should be built into daily practice.
- Evidence created the night before an audit has a distinctive scent.
- Audits should reveal process health, not inspire temporary theater.
- Noncompliance should be treated as improvement data, not personal failure.
- A mature organization does not fear audits because it already knows how it works.

Field Note 17: The Lessons Learned That Were Never Learned

Lessons learned meetings often produce excellent insights, sincere nodding, and a document that is never opened again.

- A lesson is not learned until it changes future behavior.
- Lessons should be assigned owners and follow-up actions.
- Repeated lessons indicate unmanaged organizational memory loss.
- Project closure should feed process improvement, not just archive regret.
- The most valuable lesson is the one prevented from becoming a tradition.

Field Note 18: The Status Report Theater

Status reports can clarify progress, risks, and decisions. They can also become weekly fiction written in the dialect of optimism.

- Green status without evidence is decorative confidence.
- A truthful yellow status is more valuable than a dishonest green one.

- Status reporting should expose risks early enough to act.
- If every project is green until the week before failure, the reporting process is broken.
- The purpose of status is not comfort; it is informed decision-making.

Field Note 19: The Change Control Traffic Jam

Change control exists to protect the project from uncontrolled disruption. But when every minor adjustment requires a ceremony, workers begin routing around the process.

- Change control should match the size, risk, and impact of the change.
- Not every change needs a parade.
- A change request should explain cost, schedule, quality, and stakeholder impact.
- Slow approval encourages unofficial workarounds.
- Good change control protects stability while allowing necessary movement.

Field Note 20: The Defect Priority Debate

A defect priority meeting can become a philosophical event where urgency, severity, customer embarrassment, and executive anxiety all fight for the same chair.

- Severity should describe impact; priority should describe order of action.
- Defect decisions need clear criteria before emotions enter the room.
- A defect is not less real because it is inconvenient.
- Repeated defect categories reveal process weaknesses upstream.
- Quality improves when defects are treated as information, not ammunition.

Field Note 21: The Approval Email Burial Ground

Some approvals are not lost; they are simply buried in an email thread with 47 replies, three attachments, and a subject line that no longer describes the issue.

- Email approval should be captured in the official project record.
- A decision hidden in email is not easily auditable.
- Important approvals need context, date, authority, and scope.
- If no one can find the approval, the approval may as well be folklore.
- Process memory should not depend on someone's inbox search skills.

Field Note 22: The Training Nobody Attended

Training is scheduled, announced, postponed, rescheduled, attended by half the required people, and later referenced as though universal understanding occurred.

- Training completion should be tracked, not assumed.
- People cannot follow a process they were never taught.
- Training should explain why the process exists, not just where the template lives.
- Refresher training matters when processes change.
- A trained workforce is a process asset, not an administrative checkbox.

Field Note 23: The Policy That Outlived the Department

Some policies survive reorganizations, system replacements, leadership changes, and the retirement of the person who wrote them. No one knows why they exist, but everyone fears deleting them.

- Policies should be reviewed on a defined cycle.
- Outdated policies create compliance confusion.

- Every policy should connect to a current business or process need.
- Retired policies should be archived clearly, not left to haunt the library.
- Governance includes knowing what to preserve and what to retire.

Field Note 24: The Project Plan That Believed in Fantasy

Some project plans are less schedules and more inspirational literature. They assume perfect requirements, full staffing, no defects, instant decisions, and a universe that apologizes before creating risk.

- A plan should reflect reality, not executive hope.
- Estimates should include review, rework, testing, approval, and transition effort.
- A plan without risk margin is a wish with dates.
- Workers should help validate the schedule before leadership announces it.
- Predictability improves when planning respects the actual process.

Field Note 25: The Risk Register Nobody Updated

The risk register is often created during project initiation and then ignored until one of the risks becomes a very loud issue with executive visibility.

- Risks should be reviewed regularly, not ceremonially.
- Every risk needs an owner, response strategy, and trigger condition.
- A risk that has occurred should be moved into issue management.
- Ignored risks become expensive surprises.
- Risk management is process Feng Shui: identify blocked energy before it breaks the furniture.

Field Note 26: The Compliance Checklist Mirage

A checklist can protect quality, but only if it represents real verification. Otherwise, it becomes a rapid sequence of checkmarks performed by someone trying to get to lunch.

- A checklist item should be observable, testable, or verifiable.
- Completion should mean evidence exists.
- Checklists should be updated when the process changes.
- A checklist that everyone signs without reading is a trust problem.
- Good checklists help workers remember what pressure makes them forget.

Field Note 27: The Executive Dashboard Illusion

The dashboard says everything is fine because the dashboard was designed to say everything is fine. Somewhere below the green indicators, real workers are quietly building workaround bridges.

- Dashboards should show leading indicators, not only polished outcomes.
- Executive visibility should include risk, friction, and process health.
- A dashboard that hides bad news delays correction.
- Metrics should be traceable to real data sources.
- The best dashboard invites better questions, not false comfort.

Field Note 28: The Process Champion With No Authority

The process champion is praised in meetings, assigned responsibility, and given no time, budget, authority, or political protection. Then everyone wonders why improvement is slow.

- A champion needs authority equal to the responsibility assigned.
- Process improvement requires time in the work plan.
- Champions should be supported by leadership, not sacrificed to resistance.

- A champion without access to decision-makers becomes a suggestion box.
- Real empowerment means giving process leaders the ability to remove obstacles.

Field Note 29: The Emergency Process Created During the Emergency

Organizations often discover the need for process after the crisis has already begun. Unfortunately, designing process while panicking tends to produce forms that look like smoke alarms.

- Emergency procedures should be defined before emergencies occur.
- Crisis-created processes often solve the immediate pain but ignore long-term fit.
- After-action reviews should convert emergency lessons into stable practice.
- Urgency does not eliminate the need for ownership and documentation.
- The best emergency process is the one tested before it is needed.

Field Note 30: The Real Worker Saves the Day Again

At the end of many troubled projects, after the slide decks, meetings, escalation calls, late approvals, and process confusion, a real worker quietly connects the missing pieces and saves the delivery.

- Real workers know where the process actually breaks.
- Improvement should begin with the people closest to the work.
- A mature process captures real-worker knowledge before it becomes heroic rescue.
- Heroics are admirable once; repeated heroics reveal process failure.
- IPI honors the real worker by turning survival knowledge into sustainable process discipline.

APPENDIX N

Final Implementation Toolkit Index

This index summarizes the restored legacy implementation toolkit and identifies how each form supports the larger IPI operating system.

Tool	Primary Owner	Used During	Evidence Produced
PCC Charter	PCC / Workgroup Owner	Find-Fix-Measure-Maintain	Approved artifact, log, checklist, or decision record
Representative Nomination	PCC / Workgroup Owner	Find-Fix-Measure-Maintain	Approved artifact, log, checklist, or decision record
Meeting Agenda	PCC / Workgroup Owner	Find-Fix-Measure-Maintain	Approved artifact, log, checklist, or decision record
Decision Log	PCC / Workgroup Owner	Find-Fix-Measure-Maintain	Approved artifact, log, checklist, or decision record
Policy Definition	PCC / Workgroup Owner	Find-Fix-Measure-Maintain	Approved artifact, log, checklist, or decision record
Process Definition	PCC / Workgroup Owner	Find-Fix-Measure-Maintain	Approved artifact, log, checklist, or decision record
Procedure Form	PCC / Workgroup Owner	Find-Fix-Measure-Maintain	Approved artifact, log, checklist, or decision record
Template Control	PCC / Workgroup Owner	Find-Fix-Measure-Maintain	Approved artifact, log, checklist, or decision record
Gap Card	PCC / Workgroup Owner	Find-Fix-Measure-Maintain	Approved artifact, log, checklist, or decision record
Pilot Plan	PCC / Workgroup Owner	Find-Fix-Measure-Maintain	Approved artifact, log, checklist, or decision record
Results Form	PCC / Workgroup Owner	Find-Fix-Measure-Maintain	Approved artifact, log, checklist, or decision record
Institutionalization Checklist	PCC / Workgroup Owner	Find-Fix-Measure-Maintain	Approved artifact, log, checklist, or decision record
Requirements Sign-Off	PCC / Workgroup Owner	Find-Fix-Measure-Maintain	Approved artifact, log, checklist, or decision record
Design Review	PCC / Workgroup Owner	Find-Fix-Measure-Maintain	Approved artifact, log, checklist, or decision record
Test Readiness	PCC / Workgroup	Find-Fix-Measure-	Approved artifact, log, checklist, or decision

Final Legacy Principle:

Process improvement should be simple enough to use, serious enough to audit, and human enough that the real workers do not feel punished for making the organization better.

Feng Shui Legacy Process Guide

The original concept aligns process roles with workplace values. This turns those values into practical operating guidance. Each value includes a process meaning, warning signs, legacy practices, and a field checklist that can be used by a Process Champions Committee or by any group trying to reduce frustration without surrendering discipline.

Unity: Team Work Through the Executive Management

Leadership does not merely announce process improvement; it participates in creating the conditions where improvement can survive.

Process Meaning

In the IPI workplace, unity is not a decorative word printed on a poster near the coffee machine. It is a process condition that must be visible in how decisions are made, how artifacts are controlled, how disagreements are resolved, and how real workers experience the day-to-day flow of work.

The Executive Management is responsible for protecting the team work dimension of the process. This does not mean the role owns every problem. It means the role makes sure the problem is no longer invisible, unnamed, or disguised as personality conflict.

Warning Signs

- Team Work is discussed only after a defect, escalation, or missed deadline.
- The team relies on memory instead of artifacts.
- Employees comply publicly but route around the process privately.
- The process exists, but no one can explain why it exists.
- Management asks for discipline only after approving disorder.

Legacy Practices That Support This Value

Create a named checkpoint for team work. This checkpoint should be light enough to use, clear enough to audit, and practical enough that real workers do not need a consultant, a certification, and a ceremonial spreadsheet to understand it.

Use the checkpoint to ask three questions: What is the expected artifact? Who owns it? What evidence shows the work is ready to move forward? These questions sound simple because good process questions usually are simple. Bad process questions require a glossary, an acronym decoder, and someone from a national consulting firm to explain why the answer costs another six months.

The symbolic color associated with this value is Green. In the legacy manuscript, color is used as a memory device: a way to make process roles easier to recognize and discuss. The color itself is not magic. The discipline behind the role is the magic, assuming the organization does not immediately form a subcommittee to rename the magic.

Process Value Checklist

Question	Answer / Evidence
Where does team work appear in the lifecycle?	
Who is accountable for protecting unity?	
What artifact proves the activity happened?	
What is the escalation path when the value is ignored?	
How will the PCC know the practice is improving the workplace?	

Field Note

When unity is healthy, the organization feels less frantic even when the work is difficult. When it is unhealthy, every project begins to resemble a fire drill conducted inside a filing cabinet. IPI does not promise a perfect workplace. It promises a better way to notice imbalance before imbalance becomes the project plan.

Balance: Neutral Through the Customer

Customer needs must be heard without allowing every request to become an unmanaged emergency.

Process Meaning

In the IPI workplace, balance is not a decorative word printed on a poster near the coffee machine. It is a process condition that must be visible in how decisions are made, how artifacts are controlled, how disagreements are resolved, and how real workers experience the day-to-day flow of work.

The Customer is responsible for protecting the neutral dimension of the process. This does not mean the role owns every problem. It means the role makes sure the problem is no longer invisible, unnamed, or disguised as personality conflict.

Warning Signs

- Neutral is discussed only after a defect, escalation, or missed deadline.
- The team relies on memory instead of artifacts.
- Employees comply publicly but route around the process privately.
- The process exists, but no one can explain why it exists.
- Management asks for discipline only after approving disorder.

Legacy Practices That Support This Value

Create a named checkpoint for neutral. This checkpoint should be light enough to use, clear enough to audit, and practical enough that real workers do not need a consultant, a certification, and a ceremonial spreadsheet to understand it.

Use the checkpoint to ask three questions: What is the expected artifact? Who owns it? What evidence shows the work is ready to move forward? These questions sound simple because good process questions usually are simple. Bad process questions require a glossary, an acronym decoder, and someone from a national consulting firm to explain why the answer costs another six months.

The symbolic color associated with this value is White. In the legacy manuscript, color is used as a memory device: a way to make process roles easier to recognize and discuss. The color itself is not magic. The discipline behind the role is the magic, assuming the organization does not immediately form a subcommittee to rename the magic.

Process Value Checklist

Question	Answer / Evidence
Where does neutral appear in the lifecycle?	
Who is accountable for protecting balance?	
What artifact proves the activity happened?	
What is the escalation path when the value is ignored?	
How will the PCC know the practice is improving the workplace?	

Field Note

When balance is healthy, the organization feels less frantic even when the work is difficult. When it is unhealthy, every project begins to resemble a fire drill conducted inside a filing cabinet. IPI does not promise a perfect workplace. It promises a better way to notice imbalance before imbalance becomes the project plan.

Prosperity: Worker Representation Through the Rotating Work Group Representative

The organization grows when real workers help shape the process instead of receiving it as a surprise.

Process Meaning

In the IPI workplace, prosperity is not a decorative word printed on a poster near the coffee machine. It is a process condition that must be visible in how decisions are made, how artifacts are controlled, how disagreements are resolved, and how real workers experience the day-to-day flow of work.

The Rotating Work Group Representative is responsible for protecting the worker representation dimension of the process. This does not mean the role owns every problem. It means the role makes sure the problem is no longer invisible, unnamed, or disguised as personality conflict.

Warning Signs

- Worker Representation is discussed only after a defect, escalation, or missed deadline.
- The team relies on memory instead of artifacts.
- Employees comply publicly but route around the process privately.
- The process exists, but no one can explain why it exists.
- Management asks for discipline only after approving disorder.

Legacy Practices That Support This Value

Create a named checkpoint for worker representation. This checkpoint should be light enough to use, clear enough to audit, and practical enough that real workers do not need a consultant, a certification, and a ceremonial spreadsheet to understand it.

Use the checkpoint to ask three questions: What is the expected artifact? Who owns it? What evidence shows the work is ready to move forward? These questions sound simple because good process questions usually are simple. Bad process questions require a glossary, an acronym decoder, and someone from a national consulting firm to explain why the answer costs another six months.

The symbolic color associated with this value is Purple. In the legacy manuscript, color is used as a memory device: a way to make process roles easier to recognize and discuss. The color itself is not magic. The discipline behind the role is the magic, assuming the organization does not immediately form a subcommittee to rename the magic.

Process Value Checklist

Question	Answer / Evidence
Where does worker representation appear in the lifecycle?	
Who is accountable for protecting prosperity?	
What artifact proves the activity happened?	
What is the escalation path when the value is ignored?	
How will the PCC know the practice is improving the workplace?	

Field Note

When prosperity is healthy, the organization feels less frantic even when the work is difficult. When it is unhealthy, every project begins to resemble a fire drill conducted inside a filing cabinet. IPI does not promise a perfect workplace. It promises a better way to notice imbalance before imbalance becomes the project plan.

Reputation: Process Training Through the Process Trainer

Quality reputation is strengthened when people understand the process before being judged for not following it.

Process Meaning

In the IPI workplace, reputation is not a decorative word printed on a poster near the coffee machine. It is a process condition that must be visible in how decisions are made, how artifacts are controlled, how disagreements are resolved, and how real workers experience the day-to-day flow of work.

The Process Trainer is responsible for protecting the process training dimension of the process. This does not mean the role owns every problem. It means the role makes sure the problem is no longer invisible, unnamed, or disguised as personality conflict.

Warning Signs

- Process Training is discussed only after a defect, escalation, or missed deadline.
- The team relies on memory instead of artifacts.
- Employees comply publicly but route around the process privately.
- The process exists, but no one can explain why it exists.
- Management asks for discipline only after approving disorder.

Legacy Practices That Support This Value

Create a named checkpoint for process training. This checkpoint should be light enough to use, clear enough to audit, and practical enough that real workers do not need a consultant, a certification, and a ceremonial spreadsheet to understand it.

Use the checkpoint to ask three questions: What is the expected artifact? Who owns it? What evidence shows the work is ready to move forward? These questions sound simple because good process questions usually are simple. Bad process questions require a glossary, an acronym decoder, and someone from a national consulting firm to explain why the answer costs another six months.

The symbolic color associated with this value is Red. In the legacy manuscript, color is used as a memory device: a way to make process roles easier to recognize and discuss. The color itself is not magic. The discipline behind the role is the magic, assuming the organization does not immediately form a subcommittee to rename the magic.

Process Value Checklist

Question	Answer / Evidence
Where does process training appear in the lifecycle?	
Who is accountable for protecting reputation?	
What artifact proves the activity happened?	
What is the escalation path when the value is ignored?	
How will the PCC know the practice is improving the workplace?	

Field Note

When reputation is healthy, the organization feels less frantic even when the work is difficult. When it is unhealthy, every project begins to resemble a fire drill conducted inside a filing cabinet. IPI does not promise a perfect workplace. It promises a better way to notice imbalance before imbalance becomes the project plan.

Partnership: Artifacts and Documentation Through the Process Librarian

A process library turns private knowledge into shared institutional memory.

Process Meaning

In the IPI workplace, partnership is not a decorative word printed on a poster near the coffee machine. It is a process condition that must be visible in how decisions are made, how artifacts are controlled, how disagreements are resolved, and how real workers experience the day-to-day flow of work.

The Process Librarian is responsible for protecting the artifacts and documentation dimension of the process. This does not mean the role owns every problem. It means the role makes sure the problem is no longer invisible, unnamed, or disguised as personality conflict.

Warning Signs

- Artifacts and Documentation is discussed only after a defect, escalation, or missed deadline.
- The team relies on memory instead of artifacts.
- Employees comply publicly but route around the process privately.
- The process exists, but no one can explain why it exists.
- Management asks for discipline only after approving disorder.

Legacy Practices That Support This Value

Create a named checkpoint for artifacts and documentation. This checkpoint should be light enough to use, clear enough to audit, and practical enough that real workers do not need a consultant, a certification, and a ceremonial spreadsheet to understand it.

Use the checkpoint to ask three questions: What is the expected artifact? Who owns it? What evidence shows the work is ready to move forward? These questions sound simple because good process questions usually are simple. Bad process questions require a glossary, an acronym decoder, and someone from a national consulting firm to explain why the answer costs another six months.

The symbolic color associated with this value is Pink. In the legacy manuscript, color is used as a memory device: a way to make process roles easier to recognize and discuss. The color itself is not magic. The discipline behind the role is the magic, assuming the organization does not immediately form a subcommittee to rename the magic.

Process Value Checklist

Question	Answer / Evidence
Where does artifacts and documentation appear in the lifecycle?	
Who is accountable for protecting partnership?	
What artifact proves the activity happened?	
What is the escalation path when the value is ignored?	
How will the PCC know the practice is improving the workplace?	

Field Note

When partnership is healthy, the organization feels less frantic even when the work is difficult. When it is unhealthy, every project begins to resemble a fire drill conducted inside a filing cabinet. IPI does not promise a perfect workplace. It promises a better way to notice imbalance before imbalance becomes the project plan.

Creativity: Implementation Initiative Through the Process Project Manager

Creative improvement comes from tailoring the method to the organization rather than forcing the organization to worship the method.

Process Meaning

In the IPI workplace, creativity is not a decorative word printed on a poster near the coffee machine. It is a process condition that must be visible in how decisions are made, how artifacts are controlled, how disagreements are resolved, and how real workers experience the day-to-day flow of work.

The Process Project Manager is responsible for protecting the implementation initiative dimension of the process. This does not mean the role owns every problem. It means the role makes sure the problem is no longer invisible, unnamed, or disguised as personality conflict.

Warning Signs

- Implementation Initiative is discussed only after a defect, escalation, or missed deadline.
- The team relies on memory instead of artifacts.
- Employees comply publicly but route around the process privately.
- The process exists, but no one can explain why it exists.
- Management asks for discipline only after approving disorder.

Legacy Practices That Support This Value

Create a named checkpoint for implementation initiative. This checkpoint should be light enough to use, clear enough to audit, and practical enough that real workers do not need a consultant, a certification, and a ceremonial spreadsheet to understand it.

Use the checkpoint to ask three questions: What is the expected artifact? Who owns it? What evidence shows the work is ready to move forward? These questions sound simple because good process questions usually are simple. Bad process questions require a glossary, an acronym decoder, and someone from a national consulting firm to explain why the answer costs another six months.

The symbolic color associated with this value is White. In the legacy manuscript, color is used as a memory device: a way to make process roles easier to recognize and discuss. The color itself is not magic. The discipline behind the role is the magic, assuming the organization does not immediately form a subcommittee to rename the magic.

Process Value Checklist

Question	Answer / Evidence
Where does implementation initiative appear in the lifecycle?	
Who is accountable for protecting creativity?	
What artifact proves the activity happened?	
What is the escalation path when the value is ignored?	
How will the PCC know the practice is improving the workplace?	

Field Note

When creativity is healthy, the organization feels less frantic even when the work is difficult. When it is unhealthy, every project begins to resemble a fire drill conducted inside a filing cabinet. IPI does not promise a perfect workplace. It promises a better way to notice imbalance before imbalance becomes the project plan.

Progress: Maps and Gap Analysis Through the Process Business Analyst

Progress becomes visible when the current state and desired state are mapped honestly.

Process Meaning

In the IPI workplace, progress is not a decorative word printed on a poster near the coffee machine. It is a process condition that must be visible in how decisions are made, how artifacts are controlled, how disagreements are resolved, and how real workers experience the day-to-day flow of work.

The Process Business Analyst is responsible for protecting the maps and gap analysis dimension of the process. This does not mean the role owns every problem. It means the role makes sure the problem is no longer invisible, unnamed, or disguised as personality conflict.

Warning Signs

- Maps and Gap Analysis is discussed only after a defect, escalation, or missed deadline.
- The team relies on memory instead of artifacts.
- Employees comply publicly but route around the process privately.
- The process exists, but no one can explain why it exists.
- Management asks for discipline only after approving disorder.

Legacy Practices That Support This Value

Create a named checkpoint for maps and gap analysis. This checkpoint should be light enough to use, clear enough to audit, and practical enough that real workers do not need a consultant, a certification, and a ceremonial spreadsheet to understand it.

Use the checkpoint to ask three questions: What is the expected artifact? Who owns it? What evidence shows the work is ready to move forward? These questions sound simple because good process questions usually are simple. Bad process questions require a glossary, an acronym decoder, and someone from a national consulting firm to explain why the answer costs another six months.

The symbolic color associated with this value is Gray. In the legacy manuscript, color is used as a memory device: a way to make process roles easier to recognize and discuss. The color itself is not magic. The discipline behind the role is the magic, assuming the organization does not immediately form a subcommittee to rename the magic.

Process Value Checklist

Question	Answer / Evidence
Where does maps and gap analysis appear in the lifecycle?	
Who is accountable for protecting progress?	
What artifact proves the activity happened?	
What is the escalation path when the value is ignored?	
How will the PCC know the practice is improving the workplace?	

Field Note

When progress is healthy, the organization feels less frantic even when the work is difficult. When it is unhealthy, every project begins to resemble a fire drill conducted inside a filing cabinet. IPI does not promise a perfect workplace. It promises a better way to notice imbalance before imbalance becomes the project plan.

Promotion: Inspection Through the Process Auditor

Inspection should recognize good practice as well as expose weak practice.

Process Meaning

In the IPI workplace, promotion is not a decorative word printed on a poster near the coffee machine. It is a process condition that must be visible in how decisions are made, how artifacts are controlled, how disagreements are resolved, and how real workers experience the day-to-day flow of work.

The Process Auditor is responsible for protecting the inspection dimension of the process. This does not mean the role owns every problem. It means the role makes sure the problem is no longer invisible, unnamed, or disguised as personality conflict.

Warning Signs

- Inspection is discussed only after a defect, escalation, or missed deadline.
- The team relies on memory instead of artifacts.
- Employees comply publicly but route around the process privately.
- The process exists, but no one can explain why it exists.
- Management asks for discipline only after approving disorder.

Legacy Practices That Support This Value

Create a named checkpoint for inspection. This checkpoint should be light enough to use, clear enough to audit, and practical enough that real workers do not need a consultant, a certification, and a ceremonial spreadsheet to understand it.

Use the checkpoint to ask three questions: What is the expected artifact? Who owns it? What evidence shows the work is ready to move forward? These questions sound simple because good process questions usually are simple. Bad process questions require a glossary, an acronym decoder, and someone from a national consulting firm to explain why the answer costs another six months.

The symbolic color associated with this value is Black. In the legacy manuscript, color is used as a memory device: a way to make process roles easier to recognize and discuss. The color itself is not magic. The discipline behind the role is the magic, assuming the organization does not immediately form a subcommittee to rename the magic.

Process Value Checklist

Question	Answer / Evidence
Where does inspection appear in the lifecycle?	
Who is accountable for protecting promotion?	
What artifact proves the activity happened?	
What is the escalation path when the value is ignored?	
How will the PCC know the practice is improving the workplace?	

Field Note

When promotion is healthy, the organization feels less frantic even when the work is difficult. When it is unhealthy, every project begins to resemble a fire drill conducted inside a filing cabinet. IPI does not promise a perfect workplace. It promises a better way to notice imbalance before imbalance becomes the project plan.

Longevity: Change Management Through the Configuration Manager

Change control protects the organization from repeatedly rebuilding what it never controlled.

Process Meaning

In the IPI workplace, longevity is not a decorative word printed on a poster near the coffee machine. It is a process condition that must be visible in how decisions are made, how artifacts are controlled, how disagreements are resolved, and how real workers experience the day-to-day flow of work.

The Configuration Manager is responsible for protecting the change management dimension of the process. This does not mean the role owns every problem. It means the role makes sure the problem is no longer invisible, unnamed, or disguised as personality conflict.

Warning Signs

- Change Management is discussed only after a defect, escalation, or missed deadline.
- The team relies on memory instead of artifacts.
- Employees comply publicly but route around the process privately.
- The process exists, but no one can explain why it exists.
- Management asks for discipline only after approving disorder.

Legacy Practices That Support This Value

Create a named checkpoint for change management. This checkpoint should be light enough to use, clear enough to audit, and practical enough that real workers do not need a consultant, a certification, and a ceremonial spreadsheet to understand it.

Use the checkpoint to ask three questions: What is the expected artifact? Who owns it? What evidence shows the work is ready to move forward? These questions sound simple because good process questions usually are simple. Bad process questions require a glossary, an acronym decoder, and someone from a national consulting firm to explain why the answer costs another six months.

The symbolic color associated with this value is Blue. In the legacy manuscript, color is used as a memory device: a way to make process roles easier to recognize and discuss. The color itself is not magic. The discipline behind the role is the magic, assuming the organization does not immediately form a subcommittee to rename the magic.

Process Value Checklist

Question	Answer / Evidence
Where does change management appear in the lifecycle?	
Who is accountable for protecting longevity?	
What artifact proves the activity happened?	
What is the escalation path when the value is ignored?	
How will the PCC know the practice is improving the workplace?	

Field Note

When longevity is healthy, the organization feels less frantic even when the work is difficult. When it is unhealthy, every project begins to resemble a fire drill conducted inside a filing cabinet. IPI does not promise a perfect workplace. It promises a better way to notice imbalance before imbalance becomes the project plan.

Health: Measurement Through the Metrics Coordinator

Healthy metrics reveal whether the process is improving the workplace or only decorating a dashboard.

Process Meaning

In the IPI workplace, health is not a decorative word printed on a poster near the coffee machine. It is a process condition that must be visible in how decisions are made, how artifacts are controlled, how disagreements are resolved, and how real workers experience the day-to-day flow of work.

The Metrics Coordinator is responsible for protecting the measurement dimension of the process. This does not mean the role owns every problem. It means the role makes sure the problem is no longer invisible, unnamed, or disguised as personality conflict.

Warning Signs

- Measurement is discussed only after a defect, escalation, or missed deadline.
- The team relies on memory instead of artifacts.
- Employees comply publicly but route around the process privately.
- The process exists, but no one can explain why it exists.
- Management asks for discipline only after approving disorder.

Legacy Practices That Support This Value

Create a named checkpoint for measurement. This checkpoint should be light enough to use, clear enough to audit, and practical enough that real workers do not need a consultant, a certification, and a ceremonial spreadsheet to understand it.

Use the checkpoint to ask three questions: What is the expected artifact? Who owns it? What evidence shows the work is ready to move forward? These questions sound simple because good process questions usually are simple. Bad process questions require a glossary, an acronym decoder, and someone from a national consulting firm to explain why the answer costs another six months.

The symbolic color associated with this value is Brown. In the legacy manuscript, color is used as a memory device: a way to make process roles easier to recognize and discuss. The color itself is not magic. The discipline behind the role is the magic, assuming the organization does not immediately form a subcommittee to rename the magic.

Process Value Checklist

Question	Answer / Evidence
Where does change management appear in the lifecycle?	
Who is accountable for protecting longevity?	
What artifact proves the activity happened?	
What is the escalation path when the value is ignored?	
How will the PCC know the practice is improving the workplace?	

Field Note

When health is healthy, the organization feels less frantic even when the work is difficult. When it is unhealthy, every project begins to resemble a fire drill conducted inside a filing cabinet. IPI does not promise a perfect workplace. It promises a better way to notice imbalance before imbalance becomes the project plan.

Holistic: Mentoring Through the IPI Advisor

Mentoring keeps the process human, practical, and connected to the people doing the real work.

Process Meaning

In the IPI workplace, wholistic is not a decorative word printed on a poster near the coffee machine. It is a process condition that must be visible in how decisions are made, how artifacts are controlled, how disagreements are resolved, and how real workers experience the day-to-day flow of work.

The IPI Advisor is responsible for protecting the mentoring dimension of the process. This does not mean the role owns every problem. It means the role makes sure the problem is no longer invisible, unnamed, or disguised as personality conflict.

Warning Signs

- Mentoring is discussed only after a defect, escalation, or missed deadline.
- The team relies on memory instead of artifacts.
- Employees comply publicly but route around the process privately.
- The process exists, but no one can explain why it exists.
- Management asks for discipline only after approving disorder.

Legacy Practices That Support This Value

Create a named checkpoint for mentoring. This checkpoint should be light enough to use, clear enough to audit, and practical enough that real workers do not need a consultant, a certification, and a ceremonial spreadsheet to understand it.

Use the checkpoint to ask three questions: What is the expected artifact? Who owns it? What evidence shows the work is ready to move forward? These questions sound simple because good process questions usually are simple. Bad process questions require a glossary, an acronym decoder, and someone from a national consulting firm to explain why the answer costs another six months.

The symbolic color associated with this value is Yellow. In the legacy manuscript, color is used as a memory device: a way to make process roles easier to recognize and discuss. The color itself is not magic. The discipline behind the role is the magic, assuming the organization does not immediately form a subcommittee to rename the magic.

Process Value Checklist

Question	Answer / Evidence
Where does change management appear in the lifecycle?	
Who is accountable for protecting longevity?	
What artifact proves the activity happened?	
What is the escalation path when the value is ignored?	
How will the PCC know the practice is improving the workplace?	

Field Note

When wholistic is healthy, the organization feels less frantic even when the work is difficult. When it is unhealthy, every project begins to resemble a fire drill conducted inside a filing cabinet. IPI does not promise a perfect workplace. It promises a better way to notice imbalance before imbalance becomes the project plan.

Legacy Forms and Operating Templates

The following forms expand the manuscript into a true working guide. They are intentionally practical, plain-language, and aligned to legacy process discipline. Each form can be copied into a controlled process library, used during a pilot, or adapted by the Process Champions Committee.

Legacy Project Intake Form

Purpose: Capture the business request, sponsor, scope boundary, initial risk, and expected SDLC path before work begins.

Use this form when a project, enhancement, or maintenance request needs a controlled entry point instead of a hallway promise or undocumented email chain.

Field	Response	Control Note
Project Name		Owner / Date
Sponsor		Owner / Date
Business Driver		Owner / Date
Requested Delivery Date		Owner / Date
Known Constraints		Owner / Date
Required Approvals		Owner / Date
Primary Risks		Owner / Date
Initial Process Classification		Owner / Date

Review Guidance

- Confirm the owner is a named role or person, not a department cloud.
- Confirm the artifact can be found by someone who did not create it.
- Confirm unresolved items have dates and escalation paths.
- Confirm the form supports a decision rather than preserving confusion.

IPI Reality Check

IPI note: Keep the legacy project intake form lean. Every field should either prevent rework, preserve knowledge, support a decision, or strengthen accountability.

Requirements Readiness Review

Purpose: Confirm that requirements are clear enough, approved enough, and stable enough to support downstream design, development, and test activity.

Use this review before baselining requirements or allowing the project to move forward on optimism alone.

Field	Response	Control Note
Business Objective		Owner / Date
In-Scope Functions		Owner / Date
Out-of-Scope Functions		Owner / Date
User Representatives		Owner / Date
Open Questions		Owner / Date
Traceability Status		Owner / Date
Sign-Off Authority		Owner / Date
Decision Date		Owner / Date

Review Guidance

- Confirm the owner is a named role or person, not a department cloud.
- Confirm the artifact can be found by someone who did not create it.
- Confirm unresolved items have dates and escalation paths.
- Confirm the form supports a decision rather than preserving confusion.

IPI Reality Check

IPI note: Keep the requirements readiness review lean. Every field should either prevent rework, preserve knowledge, support a decision, or strengthen accountability.

Stakeholder Commitment Record

Purpose: Document who must participate, what they owe, when they owe it, and how commitment will be verified.

Use this record when customer, user, management, or technical participation is required for project success.

Field	Response	Control Note
Stakeholder		Owner / Date
Role		Owner / Date
Commitment Made		Owner / Date
Date Committed		Owner / Date
Evidence Needed		Owner / Date
Follow-Up Date		Owner / Date
Status		Owner / Date
Escalation Required		Owner / Date

Review Guidance

- Confirm the owner is a named role or person, not a department cloud.
- Confirm the artifact can be found by someone who did not create it.
- Confirm unresolved items have dates and escalation paths.
- Confirm the form supports a decision rather than preserving confusion.

IPI Reality Check

IPI note: Keep the stakeholder commitment record lean. Every field should either prevent rework, preserve knowledge, support a decision, or strengthen accountability.

Process Gap Analysis Worksheet

Purpose: Identify where current work practice differs from the desired process and whether the gap is caused by missing guidance, weak ownership, or unrealistic expectations.

Use this worksheet during PCC analysis, audit review, or any improvement discussion where the team must separate symptoms from causes.

Field	Response	Control Note
Current Practice		Owner / Date
Expected Practice		Owner / Date
Gap Description		Owner / Date
Impact		Owner / Date
Root Cause		Owner / Date
Proposed Correction		Owner / Date
Owner		Owner / Date
Due Date		Owner / Date

Review Guidance

- Confirm the owner is a named role or person, not a department cloud.
- Confirm the artifact can be found by someone who did not create it.
- Confirm unresolved items have dates and escalation paths.
- Confirm the form supports a decision rather than preserving confusion.

IPI Reality Check

IPI note: Keep the process gap analysis worksheet lean. Every field should either prevent rework, preserve knowledge, support a decision, or strengthen accountability.

Pilot Process Selection Matrix

Purpose: Compare candidate improvements and select the pilot most likely to reduce frustration, rework, delay, or quality risk.

Use this matrix before launching a process pilot so the PCC does not improve the loudest problem while ignoring the most damaging one.

Field	Response	Control Note
Candidate Process		Owner / Date
Pain Level		Owner / Date
Risk Level		Owner / Date
Ease of Pilot		Owner / Date
Worker Support		Owner / Date
Measurement Available		Owner / Date
Recommended?		Owner / Date
Notes		Owner / Date

Review Guidance

- Confirm the owner is a named role or person, not a department cloud.
- Confirm the artifact can be found by someone who did not create it.
- Confirm unresolved items have dates and escalation paths.
- Confirm the form supports a decision rather than preserving confusion.

IPI Reality Check

IPI note: Keep the pilot process selection matrix lean. Every field should either prevent rework, preserve knowledge, support a decision, or strengthen accountability.

PCC Meeting Agenda Template

Purpose: Keep Process Champions Committee meetings focused on decisions, owners, issues, and measurable follow-up.

Use this agenda when the PCC needs structure without turning the meeting into a ceremony about future ceremonies.

Field	Response	Control Note
Agenda Item		Owner / Date
Purpose		Owner / Date
Owner		Owner / Date
Decision Needed		Owner / Date
Evidence Required		Owner / Date
Timebox		Owner / Date
Outcome		Owner / Date
Action Item		Owner / Date

Review Guidance

- Confirm the owner is a named role or person, not a department cloud.
- Confirm the artifact can be found by someone who did not create it.
- Confirm unresolved items have dates and escalation paths.
- Confirm the form supports a decision rather than preserving confusion.

IPI Reality Check

IPI note: Keep the pcc meeting agenda template lean. Every field should either prevent rework, preserve knowledge, support a decision, or strengthen accountability.

PCC Decision Log

Purpose: Record committee decisions, rationale, owners, due dates, and affected process assets.

Use this log whenever a process decision needs to survive beyond the meeting minutes and the memory of whoever brought the snacks.

Field	Response	Control Note
Decision ID		Owner / Date
Decision		Owner / Date
Rationale		Owner / Date
Approved By		Owner / Date
Date		Owner / Date
Impacted Groups		Owner / Date
Artifact Updated		Owner / Date
Review Date		Owner / Date

Review Guidance

- Confirm the owner is a named role or person, not a department cloud.
- Confirm the artifact can be found by someone who did not create it.
- Confirm unresolved items have dates and escalation paths.
- Confirm the form supports a decision rather than preserving confusion.

IPI Reality Check

IPI note: Keep the pcc decision log lean. Every field should either prevent rework, preserve knowledge, support a decision, or strengthen accountability.

Process Change Request

Purpose: Submit, evaluate, approve, defer, or reject changes to process assets in a controlled and traceable way.

Use this request when a process, template, checklist, standard, or control point needs formal revision.

Field	Response	Control Note
Request ID		Owner / Date
Current Process		Owner / Date
Requested Change		Owner / Date
Business Reason		Owner / Date
Risk if Not Changed		Owner / Date
Pilot Required		Owner / Date
Approver		Owner / Date
Disposition		Owner / Date

Review Guidance

- Confirm the owner is a named role or person, not a department cloud.
- Confirm the artifact can be found by someone who did not create it.
- Confirm unresolved items have dates and escalation paths.
- Confirm the form supports a decision rather than preserving confusion.

IPI Reality Check

IPI note: Keep the process change request lean. Every field should either prevent rework, preserve knowledge, support a decision, or strengthen accountability.

Waiver Request Form

Purpose: Document temporary exceptions to process requirements, including justification, risk, duration, approval, and follow-up.

Use this form when the team must bypass a process step without pretending the bypass is normal practice.

Field	Response	Control Note
Waiver ID		Owner / Date
Process Requirement		Owner / Date
Reason for Waiver		Owner / Date
Project Impact		Owner / Date
Expiration Date		Owner / Date
Compensating Control		Owner / Date
Approver		Owner / Date
Closure Evidence		Owner / Date

Review Guidance

- Confirm the owner is a named role or person, not a department cloud.
- Confirm the artifact can be found by someone who did not create it.
- Confirm unresolved items have dates and escalation paths.
- Confirm the form supports a decision rather than preserving confusion.

IPI Reality Check

IPI note: Keep the waiver request form lean. Every field should either prevent rework, preserve knowledge, support a decision, or strengthen accountability.

Lessons Learned Register

Purpose: Capture project learning and convert it into future process improvement rather than archived regret.

Use this register at project closeout, after major defects, after audit findings, or whenever the organization pays tuition to experience.

Field	Response	Control Note
Project/Release		Owner / Date
What Worked		Owner / Date
What Failed		Owner / Date
Root Cause		Owner / Date
Recommended Improvement		Owner / Date
Owner		Owner / Date
Target Date		Owner / Date
Reusable Asset		Owner / Date

Review Guidance

- Confirm the owner is a named role or person, not a department cloud.
- Confirm the artifact can be found by someone who did not create it.
- Confirm unresolved items have dates and escalation paths.
- Confirm the form supports a decision rather than preserving confusion.

IPI Reality Check

IPI note: Keep the lessons learned register lean. Every field should either prevent rework, preserve knowledge, support a decision, or strengthen accountability.

Audit Planning Form

Purpose: Define the audit scope, criteria, participants, evidence sources, schedule, and expected outcome before the audit begins.

Use this form when preparing a review that should reveal process truth rather than create last-minute documentation theater.

Field	Response	Control Note
Audit Scope		Owner / Date
Audit Objective		Owner / Date
Process Area		Owner / Date
Evidence List		Owner / Date
Interviewees		Owner / Date
Schedule		Owner / Date
Auditor		Owner / Date
Reporting Date		Owner / Date

Review Guidance

- Confirm the owner is a named role or person, not a department cloud.
- Confirm the artifact can be found by someone who did not create it.
- Confirm unresolved items have dates and escalation paths.
- Confirm the form supports a decision rather than preserving confusion.

IPI Reality Check

IPI note: Keep the audit planning form lean. Every field should either prevent rework, preserve knowledge, support a decision, or strengthen accountability.

Audit Finding Report

Purpose: Document what was observed, what standard was not met, evidence reviewed, impact, owner, and corrective action.

Use this report when audit results need to improve the process instead of merely alarming the people who inherited it.

Field	Response	Control Note
Finding ID		Owner / Date
Requirement		Owner / Date
Observed Condition		Owner / Date
Evidence		Owner / Date
Severity		Owner / Date
Root Cause		Owner / Date
Corrective Action		Owner / Date
Closure Date		Owner / Date

Review Guidance

- Confirm the owner is a named role or person, not a department cloud.
- Confirm the artifact can be found by someone who did not create it.
- Confirm unresolved items have dates and escalation paths.
- Confirm the form supports a decision rather than preserving confusion.

IPI Reality Check

IPI note: Keep the audit finding report lean. Every field should either prevent rework, preserve knowledge, support a decision, or strengthen accountability.

Training Needs Analysis

Purpose: Determine who needs process training, what capability is missing, and which materials are required to close the gap.

Use this analysis when people are expected to follow a process they may never have been properly taught.

Field	Response	Control Note
Role		Owner / Date
Required Process Knowledge		Owner / Date
Current Skill Level		Owner / Date
Training Method		Owner / Date
Due Date		Owner / Date
Trainer		Owner / Date
Completion Evidence		Owner / Date
Effectiveness Check		Owner / Date

Review Guidance

- Confirm the owner is a named role or person, not a department cloud.
- Confirm the artifact can be found by someone who did not create it.
- Confirm unresolved items have dates and escalation paths.
- Confirm the form supports a decision rather than preserving confusion.

IPI Reality Check

IPI note: Keep the training needs analysis lean. Every field should either prevent rework, preserve knowledge, support a decision, or strengthen accountability.

Training Attendance and Competency Record

Purpose: Record attendance, competency confirmation, refresher needs, and evidence that training produced usable knowledge.

Use this record when training must be more than a sign-in sheet passed around while people check email.

Field	Response	Control Note
Employee		Owner / Date
Role		Owner / Date
Training Module		Owner / Date
Date		Owner / Date
Score/Assessment		Owner / Date
Trainer		Owner / Date
Retake Needed		Owner / Date
Notes		Owner / Date

Review Guidance

- Confirm the owner is a named role or person, not a department cloud.
- Confirm the artifact can be found by someone who did not create it.
- Confirm unresolved items have dates and escalation paths.
- Confirm the form supports a decision rather than preserving confusion.

IPI Reality Check

IPI note: Keep the training attendance and competency record lean. Every field should either prevent rework, preserve knowledge, support a decision, or strengthen accountability.

Process Library Index

Purpose: Create a controlled inventory of current process assets, owners, locations, versions, and review dates.

Use this index when the organization needs to stop treating the shared drive like a filing cabinet after an earthquake.

Field	Response	Control Note
Artifact ID		Owner / Date
Artifact Name		Owner / Date
Owner		Owner / Date
Version		Owner / Date
Effective Date		Owner / Date
Location		Owner / Date
Review Cycle		Owner / Date
Retirement Status		Owner / Date

Review Guidance

- Confirm the owner is a named role or person, not a department cloud.
- Confirm the artifact can be found by someone who did not create it.
- Confirm unresolved items have dates and escalation paths.
- Confirm the form supports a decision rather than preserving confusion.

IPI Reality Check

IPI note: Keep the process library index lean. Every field should either prevent rework, preserve knowledge, support a decision, or strengthen accountability.

Artifact Naming Standard Worksheet

Purpose: Define naming rules that make artifacts recognizable, traceable, sortable, and less likely to be confused with older versions.

Use this worksheet when documents named final, final2, and really-final have begun threatening workplace harmony.

Field	Response	Control Note
Artifact Type		Owner / Date
Naming Pattern		Owner / Date
Version Rule		Owner / Date
Owner		Owner / Date
Storage Location		Owner / Date
Approval Needed		Owner / Date
Archive Rule		Owner / Date
Example		Owner / Date

Review Guidance

- Confirm the owner is a named role or person, not a department cloud.
- Confirm the artifact can be found by someone who did not create it.
- Confirm unresolved items have dates and escalation paths.
- Confirm the form supports a decision rather than preserving confusion.

IPI Reality Check

IPI note: Keep the artifact naming standard worksheet lean. Every field should either prevent rework, preserve knowledge, support a decision, or strengthen accountability.

Metrics Definition Sheet

Purpose: Define a metric's purpose, data source, calculation, owner, threshold, review cadence, and decision use.

Use this sheet before collecting numbers so the metric has a job other than looking serious on a dashboard.

Field	Response	Control Note
Metric Name		Owner / Date
Purpose		Owner / Date
Formula		Owner / Date
Data Source		Owner / Date
Collection Frequency		Owner / Date
Owner		Owner / Date
Threshold		Owner / Date
Action Trigger		Owner / Date

Review Guidance

- Confirm the owner is a named role or person, not a department cloud.
- Confirm the artifact can be found by someone who did not create it.
- Confirm unresolved items have dates and escalation paths.
- Confirm the form supports a decision rather than preserving confusion.

IPI Reality Check

IPI note: Keep the metrics definition sheet lean. Every field should either prevent rework, preserve knowledge, support a decision, or strengthen accountability.

Metrics Review Meeting Template

Purpose: Guide a meeting that reviews trends, thresholds, causes, actions, and owners instead of admiring charts.

Use this template when metrics should trigger decisions, correction, or process learning.

Field	Response	Control Note
Metric		Owner / Date
Trend		Owner / Date
Interpretation		Owner / Date
Decision Needed		Owner / Date
Corrective Action		Owner / Date
Owner		Owner / Date
Due Date		Owner / Date
Follow-Up		Owner / Date

Review Guidance

- Confirm the owner is a named role or person, not a department cloud.
- Confirm the artifact can be found by someone who did not create it.
- Confirm unresolved items have dates and escalation paths.
- Confirm the form supports a decision rather than preserving confusion.

IPI Reality Check

IPI note: Keep the metrics review meeting template lean. Every field should either prevent rework, preserve knowledge, support a decision, or strengthen accountability.

Release Readiness Checklist

Purpose: Confirm that functional, technical, testing, documentation, support, training, and approval conditions are satisfied before release.

Use this checklist when the organization needs to know whether release readiness is real or merely scheduled.

Field	Response	Control Note
Requirement Traceability		Owner / Date
Test Completion		Owner / Date
Defects Reviewed		Owner / Date
Configuration Baseline		Owner / Date
Training Complete		Owner / Date
Support Prepared		Owner / Date
Approval Obtained		Owner / Date
Deployment Window		Owner / Date

Review Guidance

- Confirm the owner is a named role or person, not a department cloud.
- Confirm the artifact can be found by someone who did not create it.
- Confirm unresolved items have dates and escalation paths.
- Confirm the form supports a decision rather than preserving confusion.

IPI Reality Check

IPI note: Keep the release readiness checklist lean. Every field should either prevent rework, preserve knowledge, support a decision, or strengthen accountability.

Configuration Control Record

Purpose: Document controlled items, versions, baselines, approvals, change history, and release relationships.

Use this record when artifacts, code, environments, or release packages must be protected from accidental creativity.

Field	Response	Control Note
Item		Owner / Date
Version		Owner / Date
Change Description		Owner / Date
Requester		Owner / Date
Impact		Owner / Date
Approved By		Owner / Date
Implementation Date		Owner / Date
Rollback Plan		Owner / Date

Review Guidance

- Confirm the owner is a named role or person, not a department cloud.
- Confirm the artifact can be found by someone who did not create it.
- Confirm unresolved items have dates and escalation paths.
- Confirm the form supports a decision rather than preserving confusion.

IPI Reality Check

IPI note: Keep the configuration control record lean. Every field should either prevent rework, preserve knowledge, support a decision, or strengthen accountability.

Support Transition Checklist

Purpose: Verify that operations, help desk, documentation, training, known issues, and escalation paths are ready before support begins.

Use this checklist when a system is moving from project delivery into the everyday hands of people who will receive the phone calls.

Field	Response	Control Note
Known Issues		Owner / Date
User Guide Updated		Owner / Date
Help Desk Briefed		Owner / Date
Escalation Path		Owner / Date
Monitoring Plan		Owner / Date
SLA Defined		Owner / Date
Ownership Confirmed		Owner / Date
Post-Release Review		Owner / Date

Review Guidance

- Confirm the owner is a named role or person, not a department cloud.
- Confirm the artifact can be found by someone who did not create it.
- Confirm unresolved items have dates and escalation paths.
- Confirm the form supports a decision rather than preserving confusion.

IPI Reality Check

IPI note: Keep the support transition checklist lean. Every field should either prevent rework, preserve knowledge, support a decision, or strengthen accountability.

Customer Sign-Off Record

Purpose: Capture customer approval, conditions, open issues, scope of acceptance, and evidence reviewed.

Use this record when agreement must be more reliable than a favorable nod during a meeting.

Field	Response	Control Note
Deliverable		Owner / Date
Customer Representative		Owner / Date
Acceptance Criteria		Owner / Date
Open Items		Owner / Date
Risk Accepted		Owner / Date
Sign-Off Date		Owner / Date
Conditions		Owner / Date
Next Review		Owner / Date

Review Guidance

- Confirm the owner is a named role or person, not a department cloud.
- Confirm the artifact can be found by someone who did not create it.
- Confirm unresolved items have dates and escalation paths.
- Confirm the form supports a decision rather than preserving confusion.

IPI Reality Check

IPI note: Keep the customer sign-off record lean. Every field should either prevent rework, preserve knowledge, support a decision, or strengthen accountability.

Real Worker Frustration Log

Purpose: Collect recurring pain points from the people closest to the work and convert frustration into improvement data.

Use this log when complaints reveal process energy blockages that leadership has not yet measured.

Field	Response	Control Note
Date		Owner / Date
Frustration Observed		Owner / Date
Affected Role		Owner / Date
Process Cause		Owner / Date
Suggested Fix		Owner / Date
Pattern Seen Before?		Owner / Date
PCC Review Needed		Owner / Date
Disposition		Owner / Date

Review Guidance

- Confirm the owner is a named role or person, not a department cloud.
- Confirm the artifact can be found by someone who did not create it.
- Confirm unresolved items have dates and escalation paths.
- Confirm the form supports a decision rather than preserving confusion.

IPI Reality Check

IPI note: Keep the real worker frustration log lean. Every field should either prevent rework, preserve knowledge, support a decision, or strengthen accountability.

Process Health Scorecard

Purpose: Assess whether a process is clear, used, owned, measured, trained, audited, and improving.

Use this scorecard when the PCC needs a plain-language view of process health without commissioning a consultant thesis.

Field	Response	Control Note
Process Area		Owner / Date
Clarity		Owner / Date
Usability		Owner / Date
Compliance		Owner / Date
Worker Confidence		Owner / Date
Customer Impact		Owner / Date
Metric Evidence		Owner / Date
Improvement Priority		Owner / Date

Review Guidance

- Confirm the owner is a named role or person, not a department cloud.
- Confirm the artifact can be found by someone who did not create it.
- Confirm unresolved items have dates and escalation paths.
- Confirm the form supports a decision rather than preserving confusion.

IPI Reality Check

IPI note: Keep the process health scorecard lean. Every field should either prevent rework, preserve knowledge, support a decision, or strengthen accountability.

Implementation Readiness Assessment

Purpose: Evaluate whether people, process assets, training, communication, metrics, and support are ready before rollout.

Use this assessment before declaring a process live simply because the announcement email has been drafted.

Field	Response	Control Note
Sponsor Support		Owner / Date
Worker Representation		Owner / Date
Artifact Baseline		Owner / Date
Training Plan		Owner / Date
Pilot Candidate		Owner / Date
Measurement Plan		Owner / Date
Communication Plan		Owner / Date
Risk Level		Owner / Date

Review Guidance

- Confirm the owner is a named role or person, not a department cloud.
- Confirm the artifact can be found by someone who did not create it.
- Confirm unresolved items have dates and escalation paths.
- Confirm the form supports a decision rather than preserving confusion.

IPI Reality Check

IPI note: Keep the implementation readiness assessment lean. Every field should either prevent rework, preserve knowledge, support a decision, or strengthen accountability.

Legacy SDLC Phase Gate Review

Purpose: Verify that a phase has met entry, exit, artifact, approval, quality, and readiness conditions.

Use this review when the project needs discipline between phases instead of casual momentum dressed as progress.

Field	Response	Control Note
Phase		Owner / Date
Entry Criteria		Owner / Date
Required Artifacts		Owner / Date
Exit Criteria		Owner / Date
Approver		Owner / Date
Open Risks		Owner / Date
Decision		Owner / Date
Conditions		Owner / Date

Review Guidance

- Confirm the owner is a named role or person, not a department cloud.
- Confirm the artifact can be found by someone who did not create it.
- Confirm unresolved items have dates and escalation paths.
- Confirm the form supports a decision rather than preserving confusion.

IPI Reality Check

IPI note: Keep the legacy sdlc phase gate review lean. Every field should either prevent rework, preserve knowledge, support a decision, or strengthen accountability.

Requirements Traceability Matrix

Purpose: Connect requirements to design, build, test, acceptance, change, and release evidence.

Use this matrix when the team must prove that what was requested became what was designed, built, tested, and delivered.

Field	Response	Control Note
Requirement ID		Owner / Date
Source		Owner / Date
Design Reference		Owner / Date
Test Case		Owner / Date
Defect Link		Owner / Date
Status		Owner / Date
Owner		Owner / Date
Comments		Owner / Date

wearing business clothes.

Review Guidance

- Confirm the owner is a named role or person, not a department cloud.
- Confirm the artifact can be found by someone who did not create it.
- Confirm unresolved items have dates and escalation paths.
- Confirm the form supports a decision rather than preserving confusion.

IPI Reality Check

IPI note: Keep the requirements traceability matrix lean. Every field should either prevent rework, preserve knowledge, support a decision, or strengthen accountability.

Design Review Record

Purpose: Capture design decisions, review findings, risks, approvals, open issues, and required corrections.

Use this record when the solution structure needs review before it becomes expensive code.

Field	Response	Control Note
Design Area		Owner / Date
Reviewer		Owner / Date
Concern		Owner / Date
Decision		Owner / Date
Action Item		Owner / Date
Owner		Owner / Date
Due Date		Owner / Date
Closure Evidence		Owner / Date

Review Guidance

- Confirm the owner is a named role or person, not a department cloud.
- Confirm the artifact can be found by someone who did not create it.
- Confirm unresolved items have dates and escalation paths.
- Confirm the form supports a decision rather than preserving confusion.

IPI Reality Check

IPI note: Keep the design review record lean. Every field should either prevent rework, preserve knowledge, support a decision, or strengthen accountability.

Test Readiness Review

Purpose: Confirm test scope, data, environment, entry criteria, requirements coverage, and defect management readiness.

Use this review before testing begins so quality work is prepared, not improvised.

Field	Response	Control Note
Test Scope		Owner / Date
Environment		Owner / Date
Data Needs		Owner / Date
Entry Criteria		Owner / Date
Open Blockers		Owner / Date
Roles		Owner / Date
Schedule		Owner / Date
Approval		Owner / Date

Review Guidance

- Confirm the owner is a named role or person, not a department cloud.
- Confirm the artifact can be found by someone who did not create it.
- Confirm unresolved items have dates and escalation paths.
- Confirm the form supports a decision rather than preserving confusion.

IPI Reality Check

IPI note: Keep the test readiness review lean. Every field should either prevent rework, preserve knowledge, support a decision, or strengthen accountability.

Production Deployment Approval

Purpose: Document deployment readiness, rollback plans, support contacts, approvals, and operational risks.

Use this approval when moving software into production should be a controlled transition rather than a hopeful launch.

Field	Response	Control Note
Release ID		Owner / Date
Business Approval		Owner / Date
Technical Approval		Owner / Date
Test Approval		Owner / Date
Rollback Plan		Owner / Date
Support Window		Owner / Date
Communications		Owner / Date
Final Decision		Owner / Date

Review Guidance

- Confirm the owner is a named role or person, not a department cloud.
- Confirm the artifact can be found by someone who did not create it.
- Confirm unresolved items have dates and escalation paths.
- Confirm the form supports a decision rather than preserving confusion.

IPI Reality Check

IPI note: Keep the production deployment approval lean. Every field should either prevent rework, preserve knowledge, support a decision, or strengthen accountability.

Post-Implementation Review

Purpose: Assess actual outcomes, defects, user experience, support issues, lessons, and process improvements after release.

Use this review when the organization wants to learn from delivery instead of simply surviving it.

Field	Response	Control Note
Release ID		Owner / Date
Outcome		Owner / Date
Defects After Release		Owner / Date
Customer Feedback		Owner / Date
Process Lessons		Owner / Date
Metrics		Owner / Date
Actions		Owner / Date
Owner		Owner / Date

Review Guidance

- Confirm the owner is a named role or person, not a department cloud.
- Confirm the artifact can be found by someone who did not create it.
- Confirm unresolved items have dates and escalation paths.
- Confirm the form supports a decision rather than preserving confusion.

IPI Reality Check

IPI note: Keep the post-implementation review lean. Every field should either prevent rework, preserve knowledge, support a decision, or strengthen accountability.

Executive Process Briefing Sheet

Purpose: Summarize process status, risks, decisions needed, metrics, and executive support required.

Use this sheet when leadership needs a clear view of process health without drowning in operational detail.

Field	Response	Control Note
Initiative		Owner / Date
Business Reason		Owner / Date
Current Pain		Owner / Date
Proposed Pilot		Owner / Date
Expected Benefit		Owner / Date
Risk		Owner / Date
Decision Needed		Owner / Date
Date		Owner / Date

Review Guidance

- Confirm the owner is a named role or person, not a department cloud.
- Confirm the artifact can be found by someone who did not create it.
- Confirm unresolved items have dates and escalation paths.
- Confirm the form supports a decision rather than preserving confusion.

IPI Reality Check

IPI note: Keep the executive process briefing sheet lean. Every field should either prevent rework, preserve knowledge, support a decision, or strengthen accountability.

Consultant Reality Filter

Purpose: Evaluate whether outside advice is practical, culturally aware, transferable, and useful to real workers.

Use this filter when a consulting recommendation sounds impressive but has not yet touched the actual workplace.

Field	Response	Control Note
Recommendation		Owner / Date
Evidence Provided		Owner / Date
Worker Input Included?		Owner / Date
Pilot Proposed?		Owner / Date
Cost		Owner / Date
Complexity		Owner / Date
Plain-Language Summary		Owner / Date
Accept/Reject		Owner / Date

Review Guidance

- Confirm the owner is a named role or person, not a department cloud.
- Confirm the artifact can be found by someone who did not create it.
- Confirm unresolved items have dates and escalation paths.
- Confirm the form supports a decision rather than preserving confusion.

IPI Reality Check

IPI note: Keep the consultant reality filter lean. Every field should either prevent rework, preserve knowledge, support a decision, or strengthen accountability.

Template Usefulness Review

Purpose: Determine whether a template clarifies work, captures decisions, supports evidence, and remains right-sized.

Use this review when a template begins to feel heavier than the activity it was supposed to guide.

Field	Response	Control Note
Template Name		Owner / Date
Purpose		Owner / Date
Fields Used		Owner / Date
Fields Ignored		Owner / Date
Confusing Areas		Owner / Date
Suggested Changes		Owner / Date
Owner		Owner / Date
Decision		Owner / Date

Review Guidance

- Confirm the owner is a named role or person, not a department cloud.
- Confirm the artifact can be found by someone who did not create it.
- Confirm unresolved items have dates and escalation paths.
- Confirm the form supports a decision rather than preserving confusion.

IPI Reality Check

IPI note: Keep the template usefulness review lean. Every field should either prevent rework, preserve knowledge, support a decision, or strengthen accountability.

Process Retirement Request

Purpose: Retire, archive, or replace process assets that are obsolete, duplicated, unused, or harmful.

Use this request when process cleanup requires courage, governance, and a respectful goodbye to outdated paperwork.

Field	Response	Control Note
Process/Artifact		Owner / Date
Reason to Retire		Owner / Date
Replacement		Owner / Date
Impact		Owner / Date
Archive Location		Owner / Date
Approval		Owner / Date
Effective Date		Owner / Date
Communication Needed		Owner / Date

Review Guidance

- Confirm the owner is a named role or person, not a department cloud.
- Confirm the artifact can be found by someone who did not create it.
- Confirm unresolved items have dates and escalation paths.
- Confirm the form supports a decision rather than preserving confusion.

IPI Reality Check

IPI note: Keep the process retirement request lean. Every field should either prevent rework, preserve knowledge, support a decision, or strengthen accountability.

Emergency Change Review

Purpose: Document urgent changes, risks, approvals, communication, testing impact, and post-change follow-up.

Use this review when urgency is real but the organization still needs evidence, ownership, and balance.

Field	Response	Control Note
Emergency ID		Owner / Date
Trigger		Owner / Date
Decision Time		Owner / Date
Approver		Owner / Date
Risk Accepted		Owner / Date
Post-Review Date		Owner / Date
Permanent Change Needed?		Owner / Date
Notes		Owner / Date

Review Guidance

- Confirm the owner is a named role or person, not a department cloud.
- Confirm the artifact can be found by someone who did not create it.
- Confirm unresolved items have dates and escalation paths.
- Confirm the form supports a decision rather than preserving confusion.

IPI Reality Check

IPI note: Keep the emergency change review lean. Every field should either prevent rework, preserve knowledge, support a decision, or strengthen accountability.

Process Communication Plan

Purpose: Define what process information must be communicated, to whom, when, how, and with what confirmation.

Use this plan when a process change must reach real workers before confusion reaches production.

Field	Response	Control Note
Audience		Owner / Date
Message		Owner / Date
Timing		Owner / Date
Channel		Owner / Date
Owner		Owner / Date
Feedback Method		Owner / Date
Follow-Up		Owner / Date
Completion Evidence		Owner / Date

Review Guidance

- Confirm the owner is a named role or person, not a department cloud.
- Confirm the artifact can be found by someone who did not create it.
- Confirm unresolved items have dates and escalation paths.
- Confirm the form supports a decision rather than preserving confusion.

IPI Reality Check

IPI note: Keep the process communication plan lean. Every field should either prevent rework, preserve knowledge, support a decision, or strengthen accountability.

Role Accountability Matrix

Purpose: Map roles to responsibilities, decision rights, artifacts, approvals, and escalation paths.

Use this matrix when accountability is scattered, implied, disputed, or hidden behind department names.

Field	Response	Control Note
Activity		Owner / Date
Executive		Owner / Date
Customer		Owner / Date
BA		Owner / Date
PM		Owner / Date
Developer		Owner / Date
Tester		Owner / Date
Support		Owner / Date

Review Guidance

- Confirm the owner is a named role or person, not a department cloud.
- Confirm the artifact can be found by someone who did not create it.
- Confirm unresolved items have dates and escalation paths.
- Confirm the form supports a decision rather than preserving confusion.

Process Escalation Ladder

Purpose: Define escalation levels, triggers, decision authority, timing, and evidence required.

Use this ladder when unresolved process issues need a visible path upward instead of becoming hallway folklore.

Field	Response	Control Note
Issue Type		Owner / Date
First Contact		Owner / Date
Second Level		Owner / Date

Field	Response	Control Note
Executive Path		Owner / Date
Response Time		Owner / Date
Evidence Needed		Owner / Date
Closure Rule		Owner / Date
Notes		Owner / Date

Review Guidance

- Confirm the owner is a named role or person, not a department cloud.
- Confirm the artifact can be found by someone who did not create it.
- Confirm unresolved items have dates and escalation paths.
- Confirm the form supports a decision rather than preserving confusion.

Repository Cleanup Plan

Purpose: Plan the removal, consolidation, renaming, archiving, and ownership assignment of process assets.

Use this plan when the process library needs restoration, not another folder named miscellaneous.

Field	Response	Control Note
Location		Owner / Date
Artifact Type		Owner / Date
Duplicate?		Owner / Date
Owner		Owner / Date
Keep/Archive/Delete		Owner / Date
Target Date		Owner / Date
Risk		Owner / Date
Completion Evidence		Owner / Date

Review Guidance

- Confirm the owner is a named role or person, not a department cloud.
- Confirm the artifact can be found by someone who did not create it.
- Confirm unresolved items have dates and escalation paths.
- Confirm the form supports a decision rather than preserving confusion.

Case Studies from the Process Trenches

Case Study 1 - The Requirement That Lived in a Hallway Situation

A hallway conversation became the real source of truth while the approved requirement document remained technically correct and operationally useless.

The organization believed it had a requirements process because a document existed somewhere in the process library. Unfortunately, the living process had migrated into conversations, assumptions, and individual memory. This is the moment when a legacy organization discovers that the process library is not a museum; it is supposed to be the map.

What Went Wrong

- The responsible role was assumed rather than named.
- The controlling artifact existed but was not treated as the source of truth.
- Approvals focused on dates instead of readiness.
- Risks were discussed informally and recorded too late.
- The downstream group inherited decisions it did not participate in making.

IPI Response

The Process Champions Committee should reconstruct the flow from the first decision to the final defect, identify where accountability became invisible, and define a corrective control that is lighter than bureaucracy but stronger than wishful thinking.

Step	IPI Action	Output
Find	Interview the affected roles and collect the actual artifacts used.	Observed workflow map
Fix	Define the missing control point and pilot a practical correction.	Pilot process update
Measure	Track whether the same failure pattern repeats.	Baseline and post-pilot trend
Maintain	Update the process library, training, and audit checklist.	Institutionalized change

Satirical Lesson

Never confuse the existence of a document with the existence of a process. A process document that no one follows is not governance. It is office wallpaper with page numbers.

Reusable Questions

- What decision should have been visible earlier?
- What artifact should have prevented the misunderstanding?
- Which role inherited risk without authority?
- What evidence would show that the corrected process is working?
- What should be retired, simplified, or reinforced?

Case Study 2 - The Green Status Project That Was Secretly Red Situation

Every weekly report was green because no one wanted to be the first person to announce that the plan had already detached from reality.

The organization believed it had a status reporting process because a document existed somewhere in the process library. Unfortunately, the living process had migrated into conversations, assumptions, and individual memory. This is the moment when a legacy organization discovers that the process library is not a museum; it is supposed to be the map.

What Went Wrong

- The responsible role was assumed rather than named.
- The controlling artifact existed but was not treated as the source of truth.
- Approvals focused on dates instead of readiness.
- Risks were discussed informally and recorded too late.
- The downstream group inherited decisions it did not participate in making.

IPI Response

The Process Champions Committee should reconstruct the flow from the first decision to the final defect, identify where accountability became invisible, and define a corrective control that is lighter than bureaucracy but stronger than wishful thinking.

Step	IPI Action	Output
Find	Interview the affected roles and collect the actual artifacts used.	Observed workflow map
Fix	Define the missing control point and pilot a practical correction.	Pilot process update
Measure	Track whether the same failure pattern repeats.	Baseline and post-pilot trend
Maintain	Update the process library, training, and audit checklist.	Institutionalized change

Reusable Questions

- What decision should have been visible earlier?
- What artifact should have prevented the misunderstanding?
- Which role inherited risk without authority?

- What evidence would show that the corrected process is working?
- What should be retired, simplified, or reinforced?

Case Study 3 - The Consultant Binder That Nobody Used Situation

A polished methodology binder entered the department with great ceremony and left six months later as shelf architecture.

The organization believed it had a consultant methodology process because a document existed somewhere in the process library. Unfortunately, the living process had migrated into conversations, assumptions, and individual memory. This is the moment when a legacy organization discovers that the process library is not a museum; it is supposed to be the map.

What Went Wrong

- The responsible role was assumed rather than named.
- The controlling artifact existed but was not treated as the source of truth.
- Approvals focused on dates instead of readiness.
- Risks were discussed informally and recorded too late.
- The downstream group inherited decisions it did not participate in making.

IPI Response

The Process Champions Committee should reconstruct the flow from the first decision to the final defect, identify where accountability became invisible, and define a corrective control that is lighter than bureaucracy but stronger than wishful thinking.

Step	IPI Action	Output
Find	Interview the affected roles and collect the actual artifacts used.	Observed workflow map
Fix	Define the missing	Pilot process update

Step	IPI Action	Output
	control point and pilot a practical correction.	
Measure	Track whether the same failure pattern repeats.	Baseline and post-pilot trend
Maintain	Update the process library, training, and audit checklist.	Institutionalized change

Reusable Questions

- What decision should have been visible earlier?
- What artifact should have prevented the misunderstanding?
- Which role inherited risk without authority?
- What evidence would show that the corrected process is working?
- What should be retired, simplified, or reinforced?

Case Study 4 - The Release That Passed Testing but Failed Support Situation

The application worked as designed, but the support group discovered the design at the same time as the customers.

The organization believed it had a support transition process because a document existed somewhere in the process library. Unfortunately, the living process had migrated into conversations, assumptions, and individual memory. This is the moment when a legacy organization discovers that the process library is not a museum; it is supposed to be the map.

What Went Wrong

- The responsible role was assumed rather than named.
- The controlling artifact existed but was not treated as the source of truth.
- Approvals focused on dates instead of readiness.

- Risks were discussed informally and recorded too late.
- The downstream group inherited decisions it did not participate in making.

IPI Response

The Process Champions Committee should reconstruct the flow from the first decision to the final defect, identify where accountability became invisible, and define a corrective control that is lighter than bureaucracy but stronger than wishful thinking.

Step	IPI Action	Output
Find	Interview the affected roles and collect the actual artifacts used.	Observed workflow map
Fix	Define the missing control point and pilot a practical correction.	Pilot process update
Measure	Track whether the same failure pattern repeats.	Baseline and post-pilot trend
Maintain	Update the process library, training, and audit checklist.	Institutionalized change

Reusable Questions

- What decision should have been visible earlier?
- What artifact should have prevented the misunderstanding?
- Which role inherited risk without authority?
- What evidence would show that the corrected process is working?
- What should be retired, simplified, or reinforced?

Case Study 5 - The Metrics Dashboard of Decorative Numbers
Situation

The dashboard contained many numbers, several colors, and almost no decisions.

The organization believed it had a measurement process because a document existed somewhere in the process library. Unfortunately, the living process had migrated into conversations, assumptions, and individual memory. This is the moment when a legacy organization discovers that the process library is not a museum; it is supposed to be the map.

What Went Wrong

- The responsible role was assumed rather than named.
- The controlling artifact existed but was not treated as the source of truth.
- Approvals focused on dates instead of readiness.
- Risks were discussed informally and recorded too late.
- The downstream group inherited decisions it did not participate in making.

IPI Response

The Process Champions Committee should reconstruct the flow from the first decision to the final defect, identify where accountability became invisible, and define a corrective control that is lighter than bureaucracy but stronger than wishful thinking.

Step	IPI Action	Output
Find	Interview the affected roles and collect the actual artifacts used.	Observed workflow map
Fix	Define the missing control point and pilot a practical correction.	Pilot process update
Measure	Track whether the same failure pattern repeats.	Baseline and post-pilot trend
Maintain	Update the process library, training, and audit checklist.	Institutionalized change

Reusable Questions

- What decision should have been visible earlier?
- What artifact should have prevented the misunderstanding?
- Which role inherited risk without authority?
- What evidence would show that the corrected process is working?
- What should be retired, simplified, or reinforced?

Case Study 6 - The Waiver That Became the Process Situation

An exception was granted so often that the exception became the true process and the documented process became historical fiction.

The organization believed it had a waivers process because a document existed somewhere in the process library. Unfortunately, the living process had migrated into conversations, assumptions, and individual memory. This is the moment when a legacy organization discovers that the process library is not a museum; it is supposed to be the map.

What Went Wrong

- The responsible role was assumed rather than named.

- The controlling artifact existed but was not treated as the source of truth.
- Approvals focused on dates instead of readiness.
- Risks were discussed informally and recorded too late.
- The downstream group inherited decisions it did not participate in making.

IPI Response

The Process Champions Committee should reconstruct the flow from the first decision to the final defect, identify where accountability became invisible, and define a corrective control that is lighter than bureaucracy but stronger than wishful thinking.

Step	IPI Action	Output
Find	Interview the affected roles and collect the actual artifacts used.	Observed workflow map
Fix	Define the missing control point and pilot a practical correction.	Pilot process update
Measure	Track whether the same failure pattern repeats.	Baseline and post-pilot trend
Maintain	Update the process library, training, and audit checklist.	Institutionalized change

Reusable Questions

- What decision should have been visible earlier?
- What artifact should have prevented the misunderstanding?
- Which role inherited risk without authority?
- What evidence would show that the corrected process is working?
- What should be retired, simplified, or reinforced?

Glossary of Legacy IPI Terms and Satirical Field Language

Accountability Door

The point in a process where ownership must be visible before work moves forward.

Field Use: Use this term during PCC discussions when the team needs plain language to describe a recurring pattern without turning the conversation into an accusation. Humor can lower resistance, but the corrective action still has to be real.

Artifact Archaeology

The emergency search for old documents after a project has already gone wrong.

Field Use: Use this term during PCC discussions when the team needs plain language to describe a recurring pattern without turning the conversation into an accusation. Humor can lower resistance, but the corrective action still has to be real.

Binder Gravity

The tendency of a process initiative to become heavier, slower, and less useful as more documents are added without purpose.

Field Use: Use this term during PCC discussions when the team needs plain language to describe a recurring pattern without turning the conversation into an accusation. Humor can lower resistance, but the corrective action still has to be real.

Decorative Metric

A number collected for display rather than decision-making.

Field Use: Use this term during PCC discussions when the team needs plain language to describe a recurring pattern without turning the conversation into an accusation. Humor can lower resistance, but the corrective action still has to be real.

Evidence Recovery

The act of creating or locating evidence after an audit has already begun.

Field Use: Use this term during PCC discussions when the team needs plain language to describe a recurring pattern without turning the conversation into an accusation. Humor can lower resistance, but the corrective action still has to be real.

Final Final Syndrome

The uncontrolled multiplication of document versions with names that reveal fear rather than governance.

Field Use: Use this term during PCC discussions when the team needs plain language to describe a recurring pattern without turning the conversation into an accusation. Humor can lower resistance, but the corrective action still has to be real.

Green Status Fog

The management condition where everything appears fine until the release date gets close enough to smell smoke.

Field Use: Use this term during PCC discussions when the team needs plain language to describe a recurring pattern without turning the conversation into an accusation. Humor can lower resistance, but the corrective action still has to be real.

Hallway Requirement

A requirement communicated informally and then treated as if it were formally approved.

Field Use: Use this term during PCC discussions when the team needs plain language to describe a recurring pattern without turning the conversation into an accusation. Humor can lower resistance, but the corrective action still has to be real.

Methodology Theater

Visible process activity performed primarily to look disciplined.

Field Use: Use this term during PCC discussions when the team needs plain language to describe a recurring pattern without turning the conversation into an accusation. Humor can lower resistance, but the corrective action still has to be real.

Process Interior Design

The arrangement of roles, artifacts, timing, and decisions so the workplace can flow without constant collision.

Field Use: Use this term during PCC discussions when the team needs plain language to describe a recurring pattern without turning the conversation into an accusation. Humor can lower resistance, but the corrective action still has to be real.

Real Worker Test

A practical test of whether the people who do the work can explain and use the process.

Field Use: Use this term during PCC discussions when the team needs plain language to describe a recurring pattern without turning the conversation into an accusation. Humor can lower resistance, but the corrective action still has to be real.

Template Fatigue

The condition caused by excessive forms that do not reduce risk or improve decisions.

Field Use: Use this term during PCC discussions when the team needs plain language to describe a recurring pattern without turning the conversation into an accusation. Humor can lower resistance, but the corrective action still has to be real.

Waiver Drift

The slow transformation of exceptions into normal operating procedure.

Field Use: Use this term during PCC discussions when the team needs plain language to describe a recurring pattern without turning the conversation into an accusation. Humor can lower resistance, but the corrective action still has to be real.

Workflow Sofa

A humorous metaphor for a process object blocking movement while pretending to be useful.

Field Use: Use this term during PCC discussions when the team needs plain language to describe a recurring pattern without turning the conversation into an accusation. Humor can lower resistance, but the corrective action still has to be real.

Final Premium Legacy Sections

The following five sections complete the restored legacy edition by giving the reader a memorable operating model, a field-based case study, a catalog of process anti-patterns, a 90-day restoration plan, and a closing manifesto for real workers. These sections intentionally preserve the nostalgic feel of the old ways of doing things: binders, signoffs, phase gates, shared drives, cubicle diplomacy, status meetings, defect logs, audit trails, and the honest belief that a well-run software project requires order before heroics.

The intent is not to modernize the manuscript. The intent is to make the legacy concept more complete. IPI remains a workplace balance system. Feng Shui remains the comparison point. Waterfall and structured SDLC remain the operating environment. The real worker remains the hero.

IPI REALITY CHECK: A process book does not become valuable because it is thick. It becomes valuable when a tired business analyst, project manager, tester, developer, trainer, auditor, or executive can point to a page and say, "This is exactly what keeps happening in our shop, and here is how we fix it without creating another circus."

Section I

The Legacy IPI Operating Model™

Every enduring process improvement approach needs a simple picture in the reader's mind. The original IPI concept already had one: Feng Shui for balance, SDLC for timing, CMMI for structure, Agile as a limited organizational contrast, and IPI as the people-centered implementation vehicle. This section refines that picture into a usable legacy operating model without pulling the book away from its structured waterfall center.

In the old software shop, the project did not begin with a swarm of sticky notes and a motivational chant about velocity. It began with a request, a sponsor, a project manager, a business analyst, a requirements document, a design conversation, a test plan, a release schedule, a signoff, and one brave soul asking where the latest version of the template was stored. Those artifacts were not the enemy. The enemy was disorder pretending to be speed.

The Six Questions of Legacy IPI

Why are we improving?

To remove frustration, restore order, and create workplace harmony for the people doing the real work.

What are we improving?

The processes, roles, artifacts, decision points, handoffs, and accountabilities that shape the SDLC.

When do we improve?

At planned control points: initiation, requirements, design, build, test, deployment, maintenance, and lessons learned.

Who improves the process?

The Process Champions Committee with rotating workgroup representatives, not a distant committee that has never opened the defect log.

How do we improve?

Find, Fix, Measure, and Maintain: observe the gap, repair the gap, test the correction, and keep it from drifting backward.

What proves it worked?

Reduced rework, clearer handoffs, fewer audit surprises, better documentation, visible accountability, and calmer real workers.

Legacy IPI Element	Purpose in the Workplace
Feng Shui	Creates the metaphor of balance, harmony, energy, placement, and organizational flow.
Structured SDLC	Establishes the sequence of work so people know when decisions should be made.
Process Champions Committee	Creates a governance body where real workers help define usable processes.
Four-Tier Process Hierarchy	Separates policy, process, procedure, and artifact so everything is not thrown into one binder.
Find, Fix, Measure, Maintain	Converts complaints into controlled improvement actions.
Process Library	Keeps the official method visible and accessible instead of buried in someone's personal folder.
Process Audit	Checks whether the process is being followed in reality, not merely admired in theory.
Metrics Coordinator	Keeps measurement honest so data does not become decorative wallpaper for executive slides.

Feng Shui Comparison: Balance Before Motion

Feng Shui teaches that surroundings influence human energy. In IPI terms, a chaotic process environment produces chaotic work behavior. A cluttered shared drive, an undocumented approval path, a missing requirements baseline, or a test plan nobody can locate all create negative organizational energy. That energy shows up as rework, blame, overtime, sarcasm, and the sacred hallway whisper: "Did anyone tell QA this changed?"

A balanced workplace places each process object where it belongs. Policies sit above processes. Processes guide procedures. Procedures produce artifacts. Artifacts support audits. Audits support learning. Learning supports better projects. This is not mystical. It is simply the workplace equivalent of moving the filing cabinet out of the hallway so people stop tripping over it.

Unity

The project team understands that the work is connected. Requirements chaos becomes test chaos. Test chaos becomes release chaos. Release chaos becomes customer chaos.

Balance

Accountability is present without becoming punishment. Documentation is present without becoming a museum exhibit.

Prosperity

Better process protects the business by reducing preventable cost, rework, and customer disappointment.

Reputation

The organization becomes known for delivering with discipline rather than apologizing with eloquence.

Partnership

Business and IT stop treating each other like opposing kingdoms separated by a requirements document moat.

Creativity

Workers improve the process because they are close enough to the work to know where it actually breaks.

Progress

Improvement is visible, measured, and maintained rather than announced and forgotten.

Longevity

Configuration management, change control, and process memory allow the organization to survive turnover.

Health

Metrics, training, and audits keep the process from decaying into rumor.

Holistic View

The workplace is seen as a system, not a series of isolated departments blaming each other with attachments.

The Legacy IPI Operating Model

EXECUTIVE MANAGEMENT - SPONSORSHIP, PRIORITY, AUTHORITY

PROCESS CHAMPIONS COMMITTEE

Standing Roles + Rotating Workgroup Representatives + IPI Advisor

FOUR-TIER PROCESS HIERARCHY

Policy -> Process -> Procedure -> Artifact / Record

FIND -> FIX -> MEASURE -> MAINTAIN

STRUCTURED SDLC PHASES AND CONTROL POINTS

HARMONY, ACCOUNTABILITY, QUALITY

The model is intentionally simple. It can be drawn on a whiteboard in less time than it takes a consultant to locate the projector cable. Its value is that every major IPI concept has a clear place. Nothing floats. Nothing hides. Nothing becomes a mysterious initiative with a logo, a kickoff meeting, and no owner.

IPI REALITY CHECK:

If a process model cannot be explained without a forty-seven-slide deck, it may not be a model. It may be a hostage situation with bullet points.

Operating Guidance for Executive Management

In the Legacy IPI Operating Model, executive management has a defined responsibility for restoring balance to the workplace. The old way of doing things did not fail because roles existed; it failed when roles were unclear, undocumented, overruled by urgency, or reduced to titles on an organization chart. IPI asks each role to become visible again.

- Clarify the responsibility before the phase begins, not after the defect is discovered.
- Define the artifact that proves the work was performed.
- Identify the handoff point and the person accepting the handoff.
- Record exceptions openly so the waiver does not become the secret process.
- Feed lessons learned back into the process library so the organization does not pay twice for the same mistake.

Operating Guidance for Customers and Business Sponsors

In the Legacy IPI Operating Model, customers and business sponsors has a defined responsibility for restoring balance to the workplace. The old way of doing things did not fail because roles existed; it failed when roles were unclear, undocumented, overruled by urgency, or reduced to titles on an organization chart. IPI asks each role to become visible again.

- Clarify the responsibility before the phase begins, not after the defect is discovered.
- Define the artifact that proves the work was performed.
- Identify the handoff point and the person accepting the handoff.

- Record exceptions openly so the waiver does not become the secret process.
- Feed lessons learned back into the process library so the organization does not pay twice for the same mistake.

Operating Guidance for Project Management

In the Legacy IPI Operating Model, project management has a defined responsibility for restoring balance to the workplace. The old way of doing things did not fail because roles existed; it failed when roles were unclear, undocumented, overruled by urgency, or reduced to titles on an organization chart. IPI asks each role to become visible again.

- Clarify the responsibility before the phase begins, not after the defect is discovered.
- Define the artifact that proves the work was performed.
- Identify the handoff point and the person accepting the handoff.
- Record exceptions openly so the waiver does not become the secret process.
- Feed lessons learned back into the process library so the organization does not pay twice for the same mistake.

Operating Guidance for Business Analysis

In the Legacy IPI Operating Model, business analysis has a defined responsibility for restoring balance to the workplace. The old way of doing things did not fail because roles existed; it failed when roles were unclear, undocumented, overruled by urgency, or reduced to titles on an organization chart. IPI asks each role to become visible again.

- Clarify the responsibility before the phase begins, not after the defect is discovered.
- Define the artifact that proves the work was performed.
- Identify the handoff point and the person accepting the handoff.
- Record exceptions openly so the waiver does not become the secret process.

- Feed lessons learned back into the process library so the organization does not pay twice for the same mistake.

Operating Guidance for Development

In the Legacy IPI Operating Model, development has a defined responsibility for restoring balance to the workplace. The old way of doing things did not fail because roles existed; it failed when roles were unclear, undocumented, overruled by urgency, or reduced to titles on an organization chart. IPI asks each role to become visible again.

- Clarify the responsibility before the phase begins, not after the defect is discovered.
- Define the artifact that proves the work was performed.
- Identify the handoff point and the person accepting the handoff.
- Record exceptions openly so the waiver does not become the secret process.
- Feed lessons learned back into the process library so the organization does not pay twice for the same mistake.

Operating Guidance for Quality Assurance

In the Legacy IPI Operating Model, quality assurance has a defined responsibility for restoring balance to the workplace. The old way of doing things did not fail because roles existed; it failed when roles were unclear, undocumented, overruled by urgency, or reduced to titles on an organization chart. IPI asks each role to become visible again.

- Clarify the responsibility before the phase begins, not after the defect is discovered.
- Define the artifact that proves the work was performed.
- Identify the handoff point and the person accepting the handoff.
- Record exceptions openly so the waiver does not become the secret process.

- Feed lessons learned back into the process library so the organization does not pay twice for the same mistake.

Operating Guidance for Configuration Management

In the Legacy IPI Operating Model, configuration management has a defined responsibility for restoring balance to the workplace. The old way of doing things did not fail because roles existed; it failed when roles were unclear, undocumented, overruled by urgency, or reduced to titles on an organization chart. IPI asks each role to become visible again.

- Clarify the responsibility before the phase begins, not after the defect is discovered.
- Define the artifact that proves the work was performed.
- Identify the handoff point and the person accepting the handoff.
- Record exceptions openly so the waiver does not become the secret process.
- Feed lessons learned back into the process library so the organization does not pay twice for the same mistake.

Operating Guidance for Training

In the Legacy IPI Operating Model, training has a defined responsibility for restoring balance to the workplace. The old way of doing things did not fail because roles existed; it failed when roles were unclear, undocumented, overruled by urgency, or reduced to titles on an organization chart. IPI asks each role to become visible again.

- Clarify the responsibility before the phase begins, not after the defect is discovered.
- Define the artifact that proves the work was performed.
- Identify the handoff point and the person accepting the handoff.
- Record exceptions openly so the waiver does not become the secret process.

- Feed lessons learned back into the process library so the organization does not pay twice for the same mistake.

Operating Guidance for Metrics

In the Legacy IPI Operating Model, metrics has a defined responsibility for restoring balance to the workplace. The old way of doing things did not fail because roles existed; it failed when roles were unclear, undocumented, overruled by urgency, or reduced to titles on an organization chart. IPI asks each role to become visible again.

- Clarify the responsibility before the phase begins, not after the defect is discovered.
- Define the artifact that proves the work was performed.
- Identify the handoff point and the person accepting the handoff.
- Record exceptions openly so the waiver does not become the secret process.
- Feed lessons learned back into the process library so the organization does not pay twice for the same mistake.

Operating Guidance for Process Library

In the Legacy IPI Operating Model, process library has a defined responsibility for restoring balance to the workplace. The old way of doing things did not fail because roles existed; it failed when roles were unclear, undocumented, overruled by urgency, or reduced to titles on an organization chart. IPI asks each role to become visible again.

- Clarify the responsibility before the phase begins, not after the defect is discovered.
- Define the artifact that proves the work was performed.
- Identify the handoff point and the person accepting the handoff.
- Record exceptions openly so the waiver does not become the secret process.
- Feed lessons learned back into the process library so the organization does not pay twice for the same mistake.

Operating Guidance for Process Audit

In the Legacy IPI Operating Model, process audit has a defined responsibility for restoring balance to the workplace. The old way of doing things did not fail because roles existed; it failed when roles were unclear, undocumented, overruled by urgency, or reduced to titles on an organization chart. IPI asks each role to become visible again.

- Clarify the responsibility before the phase begins, not after the defect is discovered.
- Define the artifact that proves the work was performed.
- Identify the handoff point and the person accepting the handoff.
- Record exceptions openly so the waiver does not become the secret process.
- Feed lessons learned back into the process library so the organization does not pay twice for the same mistake.

Section II Case Study: From Cubicle Chaos to Controlled Delivery

This case study is written in the spirit of the original manuscript: slightly satirical, painfully familiar, and grounded in the old software development workplace where the cubicle wall was both a privacy device and a political listening post. The organization is fictionalized, but every symptom will feel recognizable to anyone who has watched a process improvement initiative arrive wearing a new acronym and leave behind three binders, two forms, and no measurable improvement.

The Organization

Heritage Systems Group was a mid-sized software department supporting internal business applications. It had project managers, business analysts, developers, testers, release coordinators, support analysts, and executives who believed in process as long as process did not interfere with the next urgent promise made in a steering committee meeting. The department had been through multiple improvement efforts. Each arrived with energy, snacks, and a conference room reservation. Each disappeared after the first impossible deadline.

Before IPI: The Negative Qi of the Software Shop

- Requirements existed in three versions: the one signed, the one emailed, and the one everyone wished had been signed.
- The project plan contained dates that had survived every missed dependency because moving the date would have made the spreadsheet less attractive.

- Testing began when development announced it was done, which traditionally meant the first build had successfully compiled on one developer's machine.
- Configuration management was treated as a clerical inconvenience until the wrong file went to production.
- Metrics were reported in colors: green for optimism, yellow for nervousness, and red only after everyone already knew the building was on fire.
- The process library was a shared drive with folder names like Final, Final_New, Final_Use_This, and Final_Use_This_Really.
- The previous consultants had interviewed everyone, created a roadmap, and then scheduled a follow-up meeting to discuss why the roadmap required additional consulting.

The real workers were not anti-process. They were anti-nonsense. They had seen enough initiatives to know the difference between improvement and pageantry. When executives announced a new process effort, employees did not resist because they loved chaos. They resisted because experience had taught them that process improvement usually meant more work for the people already doing the work and more PowerPoint for the people explaining the work.

IPI REALITY CHECK:

Resistance is often misdiagnosed. Sometimes workers resist change because they are stubborn. Sometimes they resist because they are the only people in the room who remember what happened the last five times change was announced.

The IPI Intervention

The IPI intervention began with an uncomfortable admission: the current process was not undocumented because workers were lazy. It was undocumented because the process lived in habits, inboxes, hallway agreements, tribal memory, old templates, and the nervous laughter of people who had survived production releases.

Instead of forcing the team to invent fake current-state documentation, the IPI Advisor wrote the truth on the whiteboard: "Current State: Informal, inconsistent, personality-dependent, deadline-driven, and undocumented." For the first time, the team saw honesty treated as a process artifact.

Form the Process Champions Committee

- Select standing roles with authority.
- Assign rotating workgroup representatives.
- Make the committee responsible for practical process definition, not ceremonial attendance.

Define the Four-Tier Hierarchy

- Separate policy from process.
- Separate process from procedure.
- Separate procedure from artifact.
- Stop placing every instruction into one giant document that requires a search party.

Find the Gaps

- Document the broken handoffs.
- Identify where signoffs were unclear.
- List the forms nobody used.
- Locate the metrics that were being collected but never used for decisions.

Fix by Pilot

- Select one project.
- Use the improved requirements signoff.
- Use one change-control log.
- Use a real test readiness checklist.
- Measure the improvement before mandating it everywhere.

Measure Honestly

- Count rework caused by requirement ambiguity.
- Track defects found after handoff.
- Measure late change requests.
- Record waiver frequency.

Maintain Discipline

- Update the process library.
- Train new employees.
- Audit lightly but consistently.
- Capture lessons learned before everyone forgets the pain.

After IPI: What Changed

The department did not become perfect. Perfect projects belong in motivational posters and vendor case studies. What changed was more important: the workplace became less mysterious. Workers knew what was expected. Managers knew where the process lived. Auditors knew what evidence mattered. Business sponsors knew when signoff carried consequences. The shared drive stopped looking like an archaeological dig.

Before IPI	After IPI
Undocumented handoffs	Named handoffs with accepted artifacts
Signoff as a courtesy	Signoff as an accountable decision
Metrics as decoration	Metrics tied to rework, defects, waivers, and readiness
Process owned by outsiders	Process owned by PCC and real-worker representatives
Training by rumor	Training by role, artifact, and control point
Change control avoided	Change control understood as protection
Lessons learned forgotten	Lessons learned fed back into the library

Feng Shui Interpretation of the Case

From a Feng Shui comparison, Heritage Systems Group had bad process placement. Decisions were being made in the wrong rooms, by the wrong people, at the wrong time, without the right artifacts. Accountability was not absent; it was misplaced. Too much pressure lived at the end of the lifecycle, where testers, release coordinators, and support staff were forced to absorb decisions that should have been balanced earlier.

IPI restored the flow. Requirements energy moved into requirements review. Design energy moved into design review. Testing energy moved into test readiness. Release energy moved into change control. Learning energy moved into lessons learned. The organization did not need incense in the server room. It needed process objects placed where they belonged.

Field Note 1: A Lesson from the Case

The first process improvement is often telling the truth about the current process.

IPI REALITY CHECK:

If the hallway conversation is more accurate than the official status report, your process has already left the building.

Field Note 2: A Lesson from the Case

Workers will support process when the process protects them from preventable chaos.

Field Note 3: A Lesson from the Case

A pilot is more persuasive than a mandate because results argue better than slogans.

Field Note 4: A Lesson from the Case

A form is not bureaucracy when it captures a decision that would otherwise become a future argument.

Field Note 5: A Lesson from the Case

The process library is only useful when the official version is easier to find than the unofficial version.

Field Note 6: A Lesson from the Case

Executives do not need every detail, but they must understand the consequence of bypassing control points.

Field Note 7: A Lesson from the Case

Harmony is not the absence of conflict; it is the presence of an agreed way to resolve conflict.

Section III Process Anti-Patterns: How Good Intentions Become Bad Bureaucracy

A process anti-pattern is a repeated behavior that appears organized but creates disorder. In the legacy software environment, anti-patterns often wore respectable clothing: templates, committees, approvals, binders, dashboards, and maturity language. IPI does not reject these tools. IPI rejects their misuse. A binder can preserve knowledge or become a shrine. A metric can reveal truth or decorate a meeting. A signoff can clarify accountability or become a signature collected after the decision no longer matters.

The Binder Shrine

What it looks like: A large binder is created, placed on a shelf, and admired during audits but ignored during daily work.

Why it happens: The team confuses documentation possession with process institutionalization.

Why it damages real workers: Real workers are expected to comply with a process they cannot practically use.

How IPI corrects it: Reduce the binder into a clear hierarchy: policy, process, procedure, artifact. Keep the official version accessible and assign ownership.

Feng Shui comparison: This anti-pattern places process energy in the wrong location. It creates negative Qi because the symbol of control exists, but the actual flow of work remains blocked.

IPI REALITY CHECK:

> The Binder Shrine is dangerous because it gives management the comfort of structure without giving workers the benefit of order.

The Phantom Process

What it looks like: Everyone says there is a process, but no two people describe it the same way.

Why it happens: The process lives in memory, emails, and personalities.

Why it damages real workers: New workers inherit confusion and veterans become bottlenecks.

How IPI corrects it: Write the actual process, not the imagined one. Validate it with the people who perform it.

Feng Shui comparison: This anti-pattern places process energy in the wrong location. It creates negative Qi because the symbol of control exists, but the actual flow of work remains blocked.

The Retroactive Requirement

What it looks like: Requirements change after design, build, or testing, but the change is treated as if it was always known.

Why it happens: Decision makers avoid the discomfort of formal change control.

Why it damages real workers: Workers are blamed for missing information that did not exist when they did the work.

How IPI corrects it: Use change requests, impact analysis, and visible waiver records.

Feng Shui comparison: This anti-pattern places process energy in the wrong location. It creates negative Qi because the symbol of control exists, but the actual flow of work remains blocked.

The Decorative Metric

What it looks like: Numbers are collected because they look impressive, not because they inform decisions.

Why it happens: Executives ask for dashboards before defining the questions they need answered.

Why it damages real workers: Workers spend time feeding reports that do not improve delivery.

How IPI corrects it: Tie each metric to a decision, threshold, owner, and improvement action.

Feng Shui comparison: This anti-pattern places process energy in the wrong location. It creates negative Qi because the symbol of control exists, but the actual flow of work remains blocked.

The Checklist Nobody Checks

What it looks like: A checklist exists, but completion is assumed, skipped, or backfilled after trouble appears.

Why it happens: The checklist is treated as a formality rather than a control point.

Why it damages real workers: Control points lose meaning and audits become archaeology.

How IPI corrects it: Make checklist completion part of the phase exit criteria and require evidence.

Feng Shui comparison: This anti-pattern places process energy in the wrong location. It creates negative Qi because the symbol of control exists, but the actual flow of work remains blocked.

The Meeting About the Meeting

What it looks like: The process requires meetings to prepare for meetings that summarize meetings.

Why it happens: No artifact or decision gate defines the purpose of the meeting.

Why it damages real workers: Workers lose time and become cynical about process improvement.

How IPI corrects it: Require each meeting to have an artifact, decision, owner, or cancellation.

Feng Shui comparison: This anti-pattern places process energy in the wrong location. It creates negative Qi because the symbol of control exists, but the actual flow of work remains blocked.

The Consultant Carousel

What it looks like: Outside consultants rotate through the organization producing assessments without cultural understanding.

Why it happens: Executives mistake external confidence for internal fit.

Why it damages real workers: Real workers are studied, interrupted, and ignored.

How IPI corrects it: Make consultants listen to the PCC and validate recommendations with pilot results.

Feng Shui comparison: This anti-pattern places process energy in the wrong location. It creates negative Qi because the symbol of control exists, but the actual flow of work remains blocked.

The Maturity Level Mirage

What it looks like: The organization chases a rating or badge while ignoring the lived condition of the workplace.

Why it happens: External recognition becomes more important than internal stability.

Why it damages real workers: Workers experience process as performance for auditors.

How IPI corrects it: Treat maturity as evidence of discipline, not the purpose of discipline.

Feng Shui comparison: This anti-pattern places process energy in the wrong location. It creates negative Qi because the symbol of control exists, but the actual flow of work remains blocked.

The Email Approval Archaeology Expedition

What it looks like: Critical decisions are buried in inboxes and reconstructed during crises.

Why it happens: Formal signoff was avoided in the name of speed.

Why it damages real workers: Time is wasted searching for evidence after the fact.

How IPI corrects it: Create a decision log and store approvals in the process library or project repository.

Feng Shui comparison: This anti-pattern places process energy in the wrong location. It creates negative Qi because the symbol of control exists, but the actual

flow of work remains blocked.

The One-Size-Fits-All SDLC

What it looks like: Every project is forced through the same process regardless of size, risk, or complexity.

Why it happens: Tailoring rules do not exist or require more effort than compliance.

Why it damages real workers: Small efforts become overburdened and large efforts remain undercontrolled.

How IPI corrects it: Define tailoring criteria by risk, complexity, customer impact, and regulatory need.

Feng Shui comparison: This anti-pattern places process energy in the wrong location. It creates negative Qi because the symbol of control exists, but the actual flow of work remains blocked.

The Heroic Overtime Ritual

What it looks like: Projects depend on late-night heroics instead of early-phase discipline.

Why it happens: Planning optimism hides process weakness until the end.

Why it damages real workers: Burnout becomes a substitute for management.

How IPI corrects it: Measure rework, late change, defect leakage, and missed readiness criteria.

Feng Shui comparison: This anti-pattern places process energy in the wrong location. It creates negative Qi because the symbol of control exists, but the actual flow of work remains blocked.

The Silent Waiver

What it looks like: Teams bypass the process quietly because the official exception path is too painful.

Why it happens: The waiver process is more bureaucratic than the process being waived.

Why it damages real workers: Leadership loses visibility into where the process does not fit reality.

How IPI corrects it: Make waivers visible, simple, accountable, and subject to periodic review.

Feng Shui comparison: This anti-pattern places process energy in the wrong location. It creates negative Qi because the symbol of control exists, but the actual flow of work remains blocked.

The Template Museum

What it looks like: Templates exist for every artifact, but none are current, simple, or consistently used.

Why it happens: Template ownership is unclear and obsolete versions remain available.

Why it damages real workers: Workers waste time choosing between artifacts instead of creating useful content.

How IPI corrects it: Assign a Process Librarian, version templates, retire obsolete forms, and train by role.

Feng Shui comparison: This anti-pattern places process energy in the wrong location. It creates negative Qi because the symbol of control exists, but the actual flow of work remains blocked.

The Phase Gate Costume Party

What it looks like: A phase gate meeting is held, but decisions were already made elsewhere.

Why it happens: The gate is ceremonial rather than operational.

Why it damages real workers: The project carries hidden risk into the next phase.

How IPI corrects it: Require entry criteria, exit criteria, evidence, decisions, and documented exceptions.

Feng Shui comparison: This anti-pattern places process energy in the wrong location. It creates negative Qi because the symbol of control exists, but the actual flow of work remains blocked.

Section IV The 90-Day Legacy IPI Restoration Plan

The 90-Day Legacy IPI Restoration Plan

The 90-Day Legacy IPI Restoration Plan is designed for the old software workplace: formal phases, management reviews, project artifacts, signoffs, shared folders, meeting minutes, defect logs, and people who know that "we will clean it up later" usually means "someone in QA will discover it at the worst possible time."

This plan does not promise transformation by slogan. It restores order in controlled increments. It begins by observing reality, then forms the governance body, defines the process hierarchy, pilots improvements, measures results, and stores knowledge where future workers can find it.

Days 1-7: Admit Reality

This period focuses on admit reality. The discipline is to keep the effort practical, visible, and rooted in the actual SDLC. Every action should produce a usable artifact, a clearer decision, or a measurable reduction in workplace friction.

- Inventory existing process documents.
- Identify official and unofficial templates.
- List active projects and current phases.
- Interview real workers about recurring friction.
- Record the current process honestly, including where no formal process exists.

IPI REALITY CHECK:

> The fastest way to destroy a 90-day process restoration is to attempt a 900-day methodology conversion during the first staff meeting.

Days 8-15: Map Workplace Energy

This period focuses on map workplace energy. The discipline is to keep the effort practical, visible, and rooted in the actual SDLC. Every action should produce a usable artifact, a clearer decision, or a measurable reduction in workplace friction.

- Identify where decisions are delayed.
- Identify where rework begins.
- Identify where handoffs fail.
- Identify where approvals are unclear.
- Create the first IPI balance map: phase, artifact, owner, decision, evidence.

Days 16-22: Form the PCC

This period focuses on form the pcc. The discipline is to keep the effort practical, visible, and rooted in the actual SDLC. Every action should produce a usable artifact, a clearer decision, or a measurable reduction in workplace friction.

- Name standing roles.
- Select rotating representatives.
- Define authority.
- Set meeting rhythm.
- Create the PCC charter and communication path to executive management.

Days 23-30: Build the Four-Tier Hierarchy

This period focuses on build the four-tier hierarchy. The discipline is to keep the effort practical, visible, and rooted in the actual SDLC. Every action should produce a usable artifact, a clearer decision, or a measurable reduction in workplace friction.

- Separate policy from process.
- Separate process from procedure.
- Attach artifacts to procedures.
- Retire duplicate templates.
- Create a process library index.

Days 31-37: Find the Highest-Friction Gaps

This period focuses on find the highest-friction gaps. The discipline is to keep the effort practical, visible, and rooted in the actual SDLC. Every action should produce a usable artifact, a clearer decision, or a measurable reduction in workplace friction.

- Score gaps by business impact.
- Score gaps by worker frustration.
- Score gaps by customer impact.
- Select no more than three pilot improvements.
- Avoid boiling the ocean in a three-ring binder.

Days 38-45: Define Pilot Fixes

This period focuses on define pilot fixes. The discipline is to keep the effort practical, visible, and rooted in the actual SDLC. Every action should produce a usable artifact, a clearer decision, or a measurable reduction in workplace friction.

- Write the pilot process clearly.
- Define who uses it.
- Define when it begins and ends.
- Define what artifact proves completion.
- Define what metric will show improvement.

Days 46-52: Pilot Requirements Control

This period focuses on pilot requirements control. The discipline is to keep the effort practical, visible, and rooted in the actual SDLC. Every action should produce a usable artifact, a clearer decision, or a measurable reduction in workplace friction.

- Use a requirements readiness checklist.
- Require signoff before design baseline.
- Log late changes.
- Track ambiguity found during design or test preparation.
- Report whether the checklist reduced rework.

Days 53-60: Pilot Test Readiness

This period focuses on pilot test readiness. The discipline is to keep the effort practical, visible, and rooted in the actual SDLC. Every action should produce a usable artifact, a clearer decision, or a measurable reduction in workplace friction.

- Confirm requirements baseline.
- Confirm environment readiness.
- Confirm defect logging rules.
- Confirm entry and exit criteria.
- Measure blocked testing time.

Days 61-67: Pilot Change Control

This period focuses on pilot change control. The discipline is to keep the effort practical, visible, and rooted in the actual SDLC. Every action should produce a usable artifact, a clearer decision, or a measurable reduction in workplace friction.

- Use a change request form.
- Require impact analysis.
- Record approval, rejection, or deferral.
- Track schedule impact.
- Track waiver frequency.

Days 68-75: Measure and Review

This period focuses on measure and review. The discipline is to keep the effort practical, visible, and rooted in the actual SDLC. Every action should produce a usable artifact, a clearer decision, or a measurable reduction in workplace friction.

- Compare pilot metrics to baseline.
- Interview users of the process.
- Identify process burden versus process benefit.
- Revise the pilot artifacts.
- Prepare the PCC recommendation.

Days 76-83: Institutionalize Carefully

This period focuses on institutionalize carefully. The discipline is to keep the effort practical, visible, and rooted in the actual SDLC. Every action should produce a usable artifact, a clearer decision, or a measurable reduction in workplace friction.

- Update process library.
- Train affected roles.
- Announce the process with examples.
- Retire obsolete versions.
- Create an audit checklist.

Days 84-90: Maintain and Prepare the Next Cycle

This period focuses on maintain and prepare the next cycle. The discipline is to keep the effort practical, visible, and rooted in the actual SDLC. Every action should produce a usable artifact, a clearer decision, or a measurable reduction in workplace friction.

- Run a light audit.
- Document lessons learned.
- Select the next improvement cycle.
- Report to executive management.

- Celebrate useful order without pretending the company has achieved enlightenment.

90-Day Plan Feng Shui Alignment

Feng Shui Principle	90-Day IPI Application
Clear the clutter	Retire duplicate templates and obsolete documents.
Place objects intentionally	Put policies, processes, procedures, and artifacts in the correct hierarchy.
Balance energy	Distribute accountability across the lifecycle instead of dumping it on the release team.
Create flow	Define handoffs, phase gates, and decision points.
Sustain harmony	Train, audit, measure, and maintain after the pilot succeeds.

90-Day Executive Report Template

- Purpose of the restoration effort:

  ____

- Process areas reviewed:

  ____

- Highest-friction gaps found:

  ____

- Pilot improvements selected:

  ____

- Measures used:

  ____

- Results observed:

  ____

- Worker feedback:

  ____

- Risks remaining:

  ____

- Decisions requested:

  ____

- Next 90-day cycle recommendation:

  ____

Section V The Legacy IPI Manifesto for Real Workers

The Legacy IPI Manifesto for Real Workers

This manifesto closes the restored legacy edition by returning the book to its original purpose: taking the frustration out of the IT software development process. The old ways were not perfect. They produced their own paperwork, politics, delays, and ceremonies. But at their best, the old ways understood something valuable: a software project needs structure, memory, accountability, and respect for the people doing the work.

We believe process should serve the people doing the work.

A process that exhausts the worker while comforting the executive has lost its purpose.

Legacy practice: write down the decision, place it in the correct repository, train the people affected, and review the result before declaring victory.

We believe documentation should clarify reality, not decorate binders.

The right document at the right time can prevent weeks of rework. The wrong document, copied from an old project and renamed, is merely office furniture with page numbers.

Legacy practice: write down the decision, place it in the correct repository, train the people affected, and review the result before declaring victory.

We believe accountability should create trust, not fear.

People should know what they own, when they own it, and what evidence shows completion.

Legacy practice: write down the decision, place it in the correct repository, train the people affected, and review the result before declaring victory.

We believe metrics should reveal truth, not impress executives.

A green dashboard that hides a red project is not a metric system. It is a mood board.

Legacy practice: write down the decision, place it in the correct repository, train the people affected, and review the result before declaring victory.

We believe consultants should listen before prescribing.

No outsider should define a process without first understanding the culture, the work, the constraints, and the people who already know where the bodies are buried in the shared drive.

Legacy practice: write down the decision, place it in the correct repository, train the people affected, and review the result before declaring victory.

We believe every organization deserves a process it can actually follow.

A beautiful process that nobody follows is less useful than a simple process that workers trust.

Legacy practice: write down the decision, place it in the correct repository, train the people affected, and review the result before declaring victory.

We believe harmony is disciplined alignment.

Harmony is not soft. Harmony is what happens when people, artifacts, decisions, measures, and timing are placed correctly.

Legacy practice: write down the decision, place it in the correct repository, train the people affected, and review the result before declaring victory.

We believe the real worker is the first customer of process improvement.

If the worker cannot use the process under real conditions, the process is not finished.

Legacy practice: write down the decision, place it in the correct repository, train the people affected, and review the result before declaring victory.

We believe legacy discipline still matters.

Phase gates, signoffs, audit trails, configuration control, and lessons learned are not relics. They are reminders that software remembers every shortcut.

Legacy practice: write down the decision, place it in the correct repository, train the people affected, and review the result before declaring victory.

We believe IPI is a practical path to workplace balance.

IPI brings Feng Shui comparison into the software workplace by arranging roles, artifacts, and decisions so energy flows toward quality instead of chaos.

Legacy practice: write down the decision, place it in the correct repository, train the people affected, and review the result before declaring victory.

The Closing Feng Shui Comparison

In the physical world, a room becomes uncomfortable when furniture blocks movement, light is misplaced, objects pile up, and nothing has a natural home. In the software workplace, the same discomfort appears when requirements are misplaced, decisions are hidden, metrics are decorative, audits are punitive, and process documents are scattered across personal drives. The worker feels the imbalance before the executive sees it on a report.

IPI restores balance by placing the work in proper relation. The customer explains need. The analyst captures and clarifies it. The project manager coordinates it. The architect and developer shape it. The tester verifies it. The configuration manager protects it. The process librarian preserves it. The auditor reviews it. The metrics coordinator measures it. The trainer teaches it. The executive sponsors it. The PCC keeps it from drifting into folklore.

That is the harmony this book protects. Not a mystical escape from deadlines. Not a fantasy of perfect projects. Not a decorative philosophy placed on top of a broken process. Harmony, in the IPI sense, is the disciplined arrangement of people, process, artifacts, and accountability so real workers can do real work with less unnecessary frustration.

IPI REALITY CHECK:

The goal is not to make software development easy. The goal is to stop making it harder than it already is.

Final Words to the Real Worker

Legacy Reminder 1: Restore Flow Before Demanding Speed

When the workplace feels chaotic, look for the misplaced process object. A missing signoff, an outdated template, an unclear owner, a hidden approval, an unmeasured waiver, or a forgotten lesson learned can create more disorder than any motivational speech can repair. IPI asks you to restore placement, restore flow, and restore balance one practical improvement at a time.

Legacy Reminder 2: A Process Without an Owner Will Drift

When nobody owns the process, the process eventually becomes a rumor. It may appear in a binder, live on a shared drive, or be mentioned during orientation, but without an accountable owner it will slowly drift away from reality. IPI reminds the real worker that every meaningful process needs a guardian — someone responsible for keeping it useful, current, understood, and aligned with the work it was created to support.

Legacy Reminder 3: Documentation Is Workplace Memory

When documentation is missing, the organization forgets how it makes decisions. Requirements become conversations, approvals become vague recollections, defects become arguments, and lessons learned become stories told only by the person who has already accepted another job. IPI treats documentation as workplace memory. Not paperwork for its own sake, but the written structure that allows a team to remember, repeat, improve, and protect the work.

Legacy Reminder 4: Signoff Means Shared Understanding

A signature should never be treated as a decorative artifact at the bottom of a page. In legacy process discipline, signoff means that the right people have reviewed the right information and understand what they are approving. IPI asks the real worker to restore meaning to approval. A signoff should confirm alignment, clarify responsibility, and prevent the ancient workplace ritual of saying, "I never agreed to that," three weeks after development has already begun.

Legacy Reminder 5: Metrics Must Tell the Truth

A metric that makes everyone comfortable but helps no one decide is not measurement; it is office decoration with decimal points. IPI asks that measurement be honest, useful, and connected to action. Defect counts, schedule variance, rework trends, review findings, waiver frequency, and audit results should reveal the health of the process. When metrics tell the truth, leaders can correct imbalance before frustration becomes failure.

Legacy Reminder 6: Templates Should Help the Worker

A template should be a helpful guide, not a punishment formatted in tables. When a template becomes harder to complete than the work it describes, the organization has confused control with burden. IPI reminds the real worker that a good template reduces ambiguity, captures essential information, and supports repeatable work. If the template does not help someone think, decide, build, test, approve, or maintain, it needs improvement.

Legacy Reminder 7: Change Control Protects Balance

Change control is not meant to stop progress. It is meant to prevent uncontrolled movement from knocking the entire project out of balance. Every change carries energy through requirements, design, code, testing, schedule, cost, training, documentation, and support. IPI asks the real worker to treat change control as workplace Feng Shui: understand what is moving, who is affected, what must be adjusted, and how harmony will be restored after the change is approved.

Legacy Reminder 8: Training Keeps the Process Alive

A process that is never taught becomes tribal knowledge. A process that is poorly taught becomes frustration. A process that is taught once and never refreshed becomes mythology. IPI reminds the real worker that training is not an administrative ceremony; it is how process knowledge moves from paper into practice. When people understand why the process exists, how to follow it, and where to find help, the workplace becomes calmer and more capable.

Legacy Reminder 9: The Process Library Is the House of Knowledge

The process library is not supposed to be a digital storage closet filled with retired templates, duplicate forms, mystery folders, and documents named "final-final-approved-v3." In a balanced IPI environment, the process library is the house of knowledge. It should be orderly, current, searchable, trusted, and maintained. When the right process asset can be found at the right time, the workplace gains clarity instead of confusion.

Legacy Reminder 10: Audits Should Reveal, Not Terrorize

An audit should not feel like a surprise inspection from the process police. Properly used, an audit is a mirror. It shows whether the process being described is the process being followed. IPI asks the real worker to view audits as a way to reveal imbalance, not assign shame. If a process is not being followed, the question should be why. The answer may expose training gaps, unrealistic templates, unclear ownership, or a process that no longer fits the work.

Legacy Reminder 11: Waivers Are Warnings

A waiver is not just permission to bypass the process. It is a signal that something in the process, schedule, project, or organization requires attention. One waiver may be reasonable. Repeated waivers reveal a pattern. IPI asks the real worker to treat waivers as diagnostic evidence. If the same exception keeps appearing, the process may be too rigid, too slow, poorly understood, or misaligned with the work. The waiver is the smoke; the imbalance is the fire.

Legacy Reminder 12: Lessons Learned Must Become Lessons Applied

A lesson learned document that is filed away and forgotten is not a lesson learned; it is a confession archived for future historians. IPI reminds the real worker that the purpose of lessons learned is to change what happens next time. If the same mistake appears across projects, the organization has not learned — it has merely documented regret. True improvement occurs when lessons are reviewed, assigned, converted into process updates, and taught before the next project repeats the same pain.

Legacy Reminder 13: Real Workers Know Where the Energy Is Blocked

The people closest to the work usually know where the process is broken long before the dashboard admits it. They know which approval always stalls, which template everyone avoids, which meeting produces no decisions, which requirement format causes confusion, and which handoff creates rework. IPI honors the real worker because real workers can feel blocked process energy before management sees the report. Their insight is not resistance; it is intelligence from the workplace floor.

Legacy Reminder 14: Discipline Without Humanity Becomes Bureaucracy

Process discipline is valuable, but discipline without humanity becomes bureaucracy. A workplace can have forms, audits, approvals, templates, metrics, and governance and still fail the people doing the work. IPI asks for balance. The process must be strong enough to create order but flexible enough to respect reality. It must guide without crushing, measure without threatening, and hold people accountable without forgetting that people — not binders — deliver the work.

Legacy Reminder 15: Harmony Is the True Measure of Improvement

The final measure of process improvement is not the thickness of the binder, the number of templates, the elegance of the flowchart, or the volume of meeting minutes produced. The true measure is whether the workplace functions with more clarity, less frustration, better accountability, stronger quality, and healthier flow. IPI connects process improvement to Feng Shui because both seek the same outcome: restored balance, purposeful placement, positive energy, and harmony among the people, the work, and the environment that holds them.

Coming Next

From Intelligent Process Improvement to Process Intelligence

The legacy principles in this book were never intended to disappear.

They were intended to mature.

The structured software development world of requirements documents, signoffs, process libraries, audits, metrics, change control boards, project plans, status meetings, and carefully guarded shared drives may feel nostalgic now, but those practices carried an important lesson: software development cannot survive on motion alone. It needs structure. It needs accountability. It needs memory. It needs balance.

That was the original promise of Intelligent Process Improvement.

IPI was created to bring harmony to the workplace by helping real workers define, improve, measure, and maintain the software development processes they were expected to follow. It was not about worshiping methodology. It was not about collecting certifications. It was not about creating binders so thick they required their own ergonomic lifting policy.

It was about restoring balance.

Just as Feng Shui teaches that a healthy environment depends on the proper flow of energy, Intelligent Process Improvement teaches that a healthy software organization depends on the proper flow of work, information, accountability, and decision-making. When the process is blocked, the workplace becomes frustrated. When accountability is missing, chaos expands. When documentation is absent, memory disappears. When metrics are decorative instead of truthful, management sees shadows instead of reality.

This book preserves those legacy principles.

The next book continues the journey.

PROCESS INTELLIGENCE: The Missing Link Between Agile, DevOps, and AI expands the original IPI foundation into the modern software delivery world. Where this book looks back at the disciplined roots of structured process improvement, the next book looks forward to the operating layer now required to align Agile execution, DevOps delivery, and AI-driven systems.

The old world taught us that process without people becomes bureaucracy.

The new world teaches us that speed without intelligence becomes drift.

Together, these two books form a complete series:

Feng Shui for Intelligent Process Improvement preserves the legacy discipline of structured software development and workplace harmony.

Process Intelligence: The Missing Link Between Agile, DevOps, and AI carries those principles forward into the modern enterprise.

One book honors where software process improvement came from.

The next book shows where it must go.

But the central belief remains unchanged:

Real improvement begins when the people doing the work are given a process that creates clarity, accountability, balance, and harmony.

That was true in the age of binders, baselines, and waterfall signoffs.

It remains true in the age of Agile, DevOps, and AI.

The tools change.

The terminology changes.

The acronyms multiply.

But the need for intelligent process never goes away.

Legacy process discipline was never the enemy of progress. It was the foundation progress kept trying to outrun.

www.ingramcontent.com/pod-product-compliance
Lightning Source LLC
LaVergne TN
LVHW070217220926
848482LV00029B/7
* 9 7 9 8 9 9 5 8 7 2 5 4 2 *